CALIFORNIA NATURAL HISTORY GUIDES

FIELD GUIDE TO MUSHROOMS
OF WESTERN NORTH AMERICA

T0257053

California Natural History Guides

Phyllis M. Faber and Bruce M. Pavlik, General Editors

Field Guide to
MUSHROOMS
of Western North America

R. Michael Davis
Robert Sommer
John A. Menge

UNIVERSITY OF CALIFORNIA PRESS

Berkeley   Los Angeles   London

University of California Press, one of the most distinguished university presses in the United States, enriches lives around the world by advancing scholarship in the humanities, social sciences, and natural sciences. Its activities are supported by the UC Press Foundation and by philanthropic contributions from individuals and institutions. For more information, visit www.ucpress.edu.

California Natural History Guide Series No. 106

University of California Press
Berkeley and Los Angeles, California

University of California Press, Ltd.
London, England

Library of Congress Cataloging-in-Publication Data

Davis, R. Michael.
   Field guide to mushrooms of western North America / R. Michael Davis, Robert Sommer, John A. Menge.
       p.   cm.—(California natural history guide series ; no. 106)
   Includes bibliographical references and index.
   ISBN 978-0-520-27107-4 (cloth : alk. paper)—ISBN 978-0-520-27108-1 (paper : alk. paper)
   1. Mushrooms—California—Identification.   2. Mushrooms—West (U.S.)—Identification.   I. Sommer, Robert, 1929–   II. Menge, John A.   III. Title.
QK617.D38   2012
579.6—dc23   2011037103

Manufactured in Singapore
25   24   23   22   21
10   9   8   7   6   5   4   3

The paper used in this publication meets the minimum requirements of ANSI/NISO Z39.48-1992 (R 1997) (Permanence of Paper).

Cover photograph: *Xeromphalina campanella*. By Ron Wolf.

The publisher gratefully acknowledges the generous
contributions to this book provided by

the Gordon and Betty Moore Fund
in Environmental Studies

and

the General Endowment Fund of the University of
California Press Foundation.

CONTENTS

PREFACE

This book continues the coverage of fungi started in the California Natural History Guides by Robert and Dorothy Orr with their 1962 book on mushrooms of the San Francisco Bay Region, their 1968 book on southern California fungi, and their 1979 *Mushrooms of Western North America.* Although our focus is mainly on California fungi (including southern California, which often receives little coverage in field guides), we have collected in other western states and include some common species found between California and the southern Rockies.

Dozens of mushroom guides are in print, so why publish another? The answers are up-to-date science, scholarship, and location. Our book reflects the latest name changes, much of them based on new DNA-based technology, and we provide an appendix with old and new scientific species names covered in this book. Two authors (R. M. D. and J. A. M.) are professional mycologists, and the third (R. S.) is a serious amateur. We hope this combination makes the book both technically accurate and readable. Following the tradition of California Natural History Guides, we wanted a book light enough to carry into the woods and sufficiently detailed to facilitate accurate identification of common mushrooms found in a variety of habitats in the West.

Following the introductory sections is a descriptive key to individual species or to a genus. More detailed keys to species of popular genera are available later in the text. In total, more than 300 common or notable mushrooms (including a few molds and slime molds) of the western United States are described and illustrated. The species included in the guide were selected on the basis of their frequency in western states, distinctive color or shape, and general interest (edibility, toxicity, unusual habitat, use in making dyes, etc.). Although many other species are described in the text, this field guide is by no means comprehensive. Probably several thousand species, many poorly known, occur in the West, and collectors should not rely on a single resource to identify specimens. We strongly recommend beginners attend organized forays with persons whose knowledge of local mushrooms is sufficient to guarantee accuracy and safety.

OF THE LOWER 48 states, those in the West have the most diverse landscape. Within their borders lie the tallest mountain (Mt. Whitney at 14,494 feet) and the lowest point (Death Valley at 282 feet below sea level). Rain forests, deserts, volcanoes, foothills, valleys, and thousands of miles of coastline create microclimates that affect the types of mushrooms found and when they appear. Fall and winter are the best seasons for mushroom hunting along the coast and in the foothills. Mushroom clubs in the Pacific Northwest and in northern California often schedule their coastal forays during October through January, months when the high mountains are covered with snow and have no hint of fungi except for the occasional tree conk. High elevations are productive for fungi after snowmelt in late spring, although fall seasons with rain and no frost can also be productive. The Rockies and high mountains in Arizona can be very productive in late August after monsoon rains. Spring is morel season in much of the West. Inland valleys are not as productive

Individual species of mushrooms are typically associated with specific habitats. The Black Morel *(Morchella elata)* sometimes fruits in large numbers in montane forests the year after a forest fire.

for fungi as coastal areas, but you can find mushrooms in many locations during the rainy season and on watered lawns throughout the year.

Uncertainty is an attraction of mushroom collecting. Fruiting is predictable but never certain. Unless you have seen something with your own eyes a few days earlier, you can never be sure of what is in the field at a given time. Mushroom fruiting depends on season, temperature, and moisture. If any of these factors is unfavorable, fungi will be sparse or nonexistent. If the mushroom season is dry, mushrooms are not likely to fruit that entire year. An early frost may have a similar effect.

Moisture is less of an issue if you forage along the fog-shrouded coast, but if you seek mushrooms inland or in the Southwest—which has two seasons, dry and wet—your mushroom hunting for much of the year will be restricted to irrigated orchards, lawns, and gardens. You may also face issues of access, which is more of a problem in some locations than in others. Private land may be posted with no trespassing signs, and park land is often off-limits. Most national forests are open to noncommercial foraging, but you may need a permit. Some mushroom clubs have made special arrangements with a local agency to allow picking. If you are willing to pay a fee, you may be interested in organized forays on private land, which typically provide meals and overnight accommodations nearby. These proprietary forays are advertised in mushroom periodicals and club newsletters. Contact information for mushroom clubs and organizations is found in the backmatter of this book (Resources).

WHAT IS A MUSHROOM?

A *mushroom* is the fruiting body, or reproductive organ, of a fungus. Fungi are a unique group of organisms distinct from animals, plants, bacteria, and protozoa. Those fungi with fruiting bodies large enough to see and touch are the focus of this field guide. Fungi with microscopic fruiting bodies such as yeasts and molds receive mention here only as they relate to mushrooms. Slime molds, which are related to simple protozoan organisms, are covered briefly in this book because they share some characteristics with fungi.

Fungi consist of masses of microscopic, interwoven, and interconnected filaments individually known as *hyphae* and in mass called a *mycelium.* The mycelium of a mushroom-

Hygrocybe acutoconica, one of the brightly colored waxy caps.

producing fungus usually remains out of sight underground or embedded in its substrate. Thus, a specific fungus may fruit year after year in the same place from the same mycelial body.

Unlike green plants, fungi do not possess chlorophyll and cannot produce their own food. Instead, they live on food originally produced by plants or animals. Threads of mycelium running through leaf litter on the forest floor, for example, form an extensive network that breaks down organic matter by excreting enzymes and then absorbing the carbohydrates, amino acids, vitamins, and other nutrients through the walls of the hyphae. Although it usually goes unnoticed, the mycelium of a single fungus can extend over acres of forest and reach several feet underground. Scientists have demonstrated that some fungi grow to gigantic proportions. In an eastern Oregon forest, a colony of *Armillaria solidipes,* which produces bundles of hyphae called *rhizomorphs,* was estimated to cover more than 2,000 acres. The weight of the colony may exceed 200 tons! If the colony is considered a single organism, then it is one of the largest if not the largest organism on Earth. When conditions are right, the fungus mycelium forms a knot that develops into a mushroom fruiting body. The fruiting body produces seedlike spores that serve as the reproductive units. A typical mushroom produces millions of spores dispersed by wind, water, insects, or animals to grow into mycelia at new locations.

Most of the fungi included in this field guide produce spores on one of two types of large, fleshy fruiting bodies. Members of the phylum Basidiomycota (loosely called basidiomycetes) typically produce four basidiospores on the outside of club-shaped microscopic cells called *basidia* (singular *basidium*). Members of the phylum Ascomycota (ascomycetes) produce ascospores, usually eight in number, inside thin, spherical to fingerlike sacs called *asci* (singular *ascus*). While the spore-bearing cells are microscopic, these two general groups of fungi often can be distinguished by the shape of the fruiting bodies. Fungi that have gills, teeth, tubes, pores, or spherical above-ground spore sacs are usually basidiomycetes, whereas fungi that are shaped like cups, saucers, or goblets are usually ascomycetes (see Figure 1).

Basidiomycetes have traditionally been classified based on the shape and structure of the fruiting body and especially the portion of the fruiting body that is lined with the basidia. Fruit-

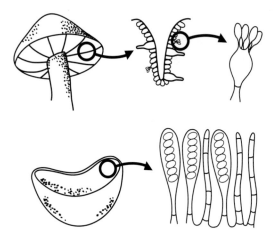

Figure 1. Members of basidiomycetes (top) bear their spores on club-shaped basidia on gills, spines, pores, and such. Ascomycetes (bottom) produce spores in saclike asci that line cups, pits, chambers, and other surfaces.

ing bodies may be erect, resupinate (lying flat against the substrate), shelflike, or effused-reflexed (partly shelflike and partly resupinate).

Genetic analyses have revealed that fruiting body shapes often are not good predictors of relationships among fungi. Nevertheless, for convenience we can categorize basidiomycetes based on their physical features for ease of identification. Thus, fungi that produce spores on platelike gills are called *gilled mushrooms,* and many that have pores or tubes, are soft and fleshy, and are shaped like mushrooms are known as *boletes.* Other fungi that are tough or woody with spore-bearing tubes or pores are called *polypores.* Fungi with downward-pointed teeth are known as *tooth fungi.* Others that bear spores on upright simple or branched fruiting bodies resembling erect clubs or corals are called *club fungi* and *coral fungi.* Fungi that produce spores inside an enclosed spherical structure are known as *puffballs* and *earthstars.*

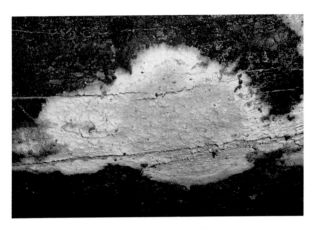

Hyphoderma praetermissum, a resupinate crust fungus.

Some ascomycetes bear their spores inside tiny, hollow, flask-shaped structures. These spore-bearing flasks are often clustered within a larger mass of rigid mycelia called a *stroma.* In the *cup fungi* the spore-bearing surface (*hymenium*) lines the inside of a cup- or saucer-shaped fruiting body. Asci are therefore directly exposed to the outside environment. In the subterranean *truffles,* the spore-bearing surface may be greatly convoluted as it lines labyrinthine-like folds inside the fruiting body. In other cases, ascus-lined cups have fused together in complex caps, as in the *morels.*

In a typical basidiomycete fruiting body, the mushroom fundamentally consists of a cap, gills, and a stalk (see Figure 2). The cap protects gills lined with the hymenium. When a mushroom is young, in the button stage, the cap often curves downward and inward against the stalk. This protects the cap against breaking as it pushes upward through the soil.

As a mushroom matures, the cap opens like an umbrella. The shape assumed by the mature cap is characteristic for a particular species. Caps may be hemispherical, convex, umbonate, conical, flattened, urn shaped, and so forth (see Figure 3), and may be smooth, pitted, wrinkled, or striate (radially lined). Stalks may be equal (with parallel sides), tapered toward the top or

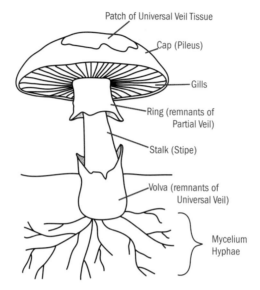

Figure 2. Mushroom features.

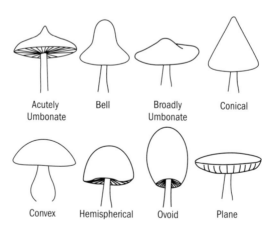

Figure 3. Cap shapes.

base, club shaped, bulbous, wiry, or hairlike, and short, long, central, eccentric (off-center), lateral (attached to the side of the cap), or absent. Some are reticulate (resembling a mesh netting stretched over the stalk).

Gills are a series of thin plates radiating from the stalk to the underside of the cap margin. The spacing between gills and their manner of attachment to the stalk are important taxonomic characteristics. Gill spacing can be divided into four categories: distant, subdistant, close, and crowded. Gill attachment to the stalk is described by the terms *free, adnate, adnexed, notched,* and *decurrent* (see Figure 4). All gilled mushrooms have full-length gills that extend from the edge of the cap to the stalk. Some species have additional short gills that extend from the cap margin and partway to the stalk.

Many mushrooms have protective structures known as veils. The universal veil or outer veil is a protective tissue that envelops the young mushrooms of some species. The veil breaks when the stalk elongates and the cap opens. Remnants of the veil left on the lower part of the stalk form a cup or volva, or may remain on the cap surface as patches or warts.

The surest way to determine if a particular species has a universal veil is to examine a young mushroom. If this is not possi-

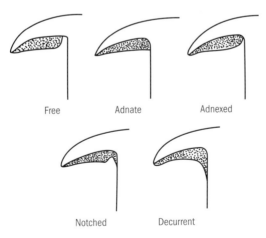

Figure 4. Types of gill attachment with the stalk.

The universal veil of *Amanita calyptroderma* is thick and cottony.

ble, the mature mushroom must be carefully examined for veil remnants. A partial veil extends from the edge of the cap to partway down the stalk, protecting the hymenium when the mushroom is young. As the mushroom cap expands, the partial veil leaves remnants on the stalk, the cap margin, or both. A stalk that has remnants of the partial veil is said to have a *ring* or *annulus*. Veils are often like a membrane or sheet (*membranous*), but sometimes they are thin and ephemeral.

Some cap surfaces are dry, whereas others are viscid or slimy. The cap must be moist to detect a viscid surface, so if a fruiting body is dry, you may need to rub water on it. Clues that a fruiting body is viscid include the presence of a shiny lacquer or adhering leaves and other debris.

Color is one of the most conspicuous characteristics of fungi and can vary tremendously within a species. *Russula cremoricolor,* for example, can be red, yellow, or pink, whereas *R. bicolor* is a combination of various shades of pink and yellow. Some of these color variations may be due to changes that take place as the fruiting body ages, but some variations are governed by genetics. The color of fruiting bodies also can be influenced by the environment. Sunlight especially can affect color: if a fruiting body grows in the shade or under leaves, it may be pale relative to its normal color; conversely, fruiting bodies that grow in full sun may become "sunburned" and develop abnormal shades of color. In addition, heavy rains may wash out colors, and freezing and thawing can radically affect mushroom appearance. In some species, the moisture in a mushroom cap affects its color. A mushroom cap that becomes lighter as moisture evaporates from the flesh is called *hygrophanous*. Some caps are partially translucent when moist, allowing the gills to show through the cap, especially near the margin where the flesh is thinnest. Color changes or staining may occur in age or when fungal tissue is bruised or cut. Injured tissue may change color almost instantaneously, or the changes may become evident after several minutes or hours.

Odor is an important and often highly specific characteristic of many fungi. Some odors attract insects and animals to aid in the dissemination of spores. Truffles, for example, use their powerful odors to attract animals that will dig them up. Stinkhorns use their strong foul scent to attract insects that will carry away their spores. Sometimes, you may need to crush a specimen to release an odor. It is best to use fresh specimens, because aging specimens often lose odors or develop odors of decay.

Like odor, taste is sometimes used to identify specific fungi. Typical tastes are classified as mild, sweet, bitter, or acrid (peppery). Only taste specimens in groups known to be nonpoisonous. When in doubt, don't taste it. To assess the taste of a fungus, place a small piece of cap and gills on the end of your tongue before spitting it out. Many tastes are slow to develop. Use only fresh specimens, because some tastes may be associated with aging or decay.

Spore color is one of the first characteristics needed for mushroom identification. To determine spore color, you need a mass of spores because a single spore is microscopic (see p. 20 for instructions on making a spore print). Other spore traits, such as size, shape, and ornamentation, are important characteristics used to identify some species but are not stressed in this field guide because they require microscopic examination. Some of the common terms used to describe spore ornamentation are *smooth, spiny, warted, ridged,* or *reticulate* (intersecting ridges resembling a honeycomb). Often the spores react to certain chemicals. By far the most useful of these chemicals is iodine (Melzer's reagent). It is naturally reddish brown in color but turns bluish in the presence of starch, called an *amyloid reaction*.

Cystidia are microscopic end cells of various shapes that often have great taxonomic importance. In some species, they are covered with crystalline material or encrusted with other substances. Others have oily or refractive inclusions. Cystidia may occur on gill edges, on the sides of gills, on the stalk, or on the cap cuticle (pileocystidia).

For field identification, there is no substitute for experience. Keep notes on mushrooms you identify and you'll soon start recognizing patterns in color, shapes, and habitats. Become familiar with basic fruiting body architecture, such as the types and variations in veils, so you can quickly categorize certain groups of species. And don't forget the importance of spore color. This guide, as well as most other guides, uses spore color as one of the first steps toward identification.

FUNGAL ECOLOGY

Most fungi can be placed into one of three ecological categories according to the way they obtain their nutritional requirements: saprobic, parasitic, and mycorrhizal. *Saprobic fungi* absorb nutrients from plant litter, wood, dung, and so on. The process of decomposition breaks down organic matter and recycles nutrients, essential functions for the web of life. How wood is decomposed is a useful taxonomic characteristic. White rot fungi decompose both the lignin and cellulose of wood, leaving behind white residual cellulose in the partially decomposed wood. Brown rot fungi, in contrast, decompose the cellulose and leave the lignin intact, resulting in a rot characterized by small brown cubes of wood. Wood rot fungi help produce humus, which is critical for soil structure and growth of plants. Some fungi break down lawn thatch, growing from a center point and fruiting in fairy rings. The lawn at the growing margin is greener than the

Fairy rings indicate the active growing margin of a colony of a fungus decomposing lawn thatch.

rest of the lawn because the fungus creates a flush of nutrients as it decomposes organic matter.

Parasitic fungi extract nutrients from living plants, resulting in plant disease and sometimes plant death. Some cause billions of dollars in losses to farmers and foresters. Notable examples are *Armillaria* root rot, the cause of extensive economic losses to ornamental trees, fruit and nut production, and the timber industry; Dutch elm disease, which has devastated elm trees across the United States; white pine blister rust; and late blight, which caused the Irish potato famine in the mid-1800s and continues to plague potato production today. Sudden oak death, caused by the funguslike organism *Phytophthora ramorum,* is a serious disease of oaks and other plants of the Pacific Northwest and California.

Mycorrhizal fungi grow in a symbiotic, mutually beneficial relationship with plants. The name *mycorrhiza* is derived from the Greek *myco,* meaning fungus, and *rhiza,* or root. The mycelium penetrates the plant's roots and forms a protective sheath around them. The mycelium radiates from the mycorrhizal roots and explores the soil. This mycelial network greatly increases the absorptive surface area of the root and mines the soil for nutrients such as phosphorus, zinc, copper, potassium, and nitrogen, which it passes to the host plant. Because of this symbiotic relationship, many trees can grow and prosper in extremely poor soils. In return, the host tree provides the fungus with carbohydrates, amino acids, and vitamins, which may be the fungus's sole food supply. Without these mycorrhizal associations, pines, firs, manzanitas, oaks, aspens, birch, and many other tree species would struggle to survive. These forests provide collectors a bounty of mycorrhizal mushrooms.

Although a single mushroom cap may release millions of spores, few spores arrive on a suitable substrate at the right time to germinate and successfully compete for nutrients. Thin-walled and nonpigmented spores are easily damaged by ultraviolet light and desiccation and cannot survive long-distance travel. Because much higher concentrations of spores arrive near the parent fungus, fungi often become somewhat localized. When spores arrive on a suitable substrate, they may germinate and produce hyphae, but only a mycelium generated from two spores of different mating types can produce fruiting bodies. Fungal mating types are not like the "male" and "female" system in animals, with only a 50 percent rate of sexual compatibility. Instead, there may be large numbers of mating types, most of which are sexually compatible. However, the spores from one fruiting body usually have limited sexual compatibility, which discourages inbreeding and promotes outbreeding, providing genetic diversity, and hence adaptive ability.

Fungi often become localized because of habitat specificity and precise requirements for moisture and temperature. Due to the arid climate of much of the western United States, a large number of western fungi have evolved mechanisms to cope with drought. Indeed, climate is a driving force for the evolution of our unique fungal flora. Truffles, for example, have developed a subterranean existence where moisture levels are more consistent and manageable. These fungi require external agents such as animals and insects to spread their spores. Other fungi have adapted to drought conditions by allowing their fruiting bodies to dry out and then quickly rehydrate during rare rain events. Some *Marasmius* species, jelly fungi, and many of the crust fungi fall into this category. Still others manage by growing under logs or remaining under the litter layer and producing "shrumps" (mushroom humps). This conserves moisture and allows spores to be released even during relatively dry periods. Many fungi, such as puffballs and bird's nest fungi (e.g., *Nidula candida*), are adapted for rain dissemination and release their spores only when abundant moisture is present. Others, such as *Podaxis pistillaris,* a common desert and chaparral fungus, produce thick-walled dark spores that are well suited to survive long periods of drought, sun, and heat.

COLLECTING MUSHROOMS

Starting Out

When you first become interested in mushrooms, the number and variety of species can be overwhelming, and learning all the names may seem daunting. The most common fungi in your area will have one or more common names (e.g., Inky Cap) and a technical name consisting of two Latin or latinized words, the first indicating genus (*Coprinopsis*) and a second for species (*atramentaria*). Rarely, a third Latin word preceded by "var." is used to indicate variety. These technical names change as we make new discoveries. For example, inky caps used to be placed in the genus *Coprinus,* not in the present genus, *Coprinopsis.* Often only the genus name changes. The species name won't change unless we learn that someone named the mushroom earlier and used a different name. Common names tend to remain stable within a location because they refer to observable features, such as color, shape, staining, or, in this case, turning into ink. However, common names can vary greatly between regions, and for this reason common names must be used with caution.

The best first step for beginners is to learn the distinctive fungi in your area. Don't try to be systematic and learn names for the amanitas or boletes that you will seldom encounter—that will come later. You can also leave the LBMs (little brown mushrooms) for another time. Often they are difficult to identify, and perhaps learning the genus or common name for the group would be adequate. Concentrate on the most common, showy, and significant species in your area. Mushrooming is a regional activity, and what is common and distinctive in one region may be uncommon or nonexistent in another. Start opportunistically to identify those genera (plural of *genus*) that are familiar and seem interesting. Their distinctive features will soon become evident—pores or gills, growing on wood or on the ground, spore color, dry or viscid cap, cap size, staining or bruising reactions, and the presence of latex, among other observable characteristics. You will be surprised at how quickly such distinctive features become recognizable.

It may be tempting to skip the nomenclature and concentrate on a few easy edible species. Some mushroomers limit their collecting in this way. This is true of many commercial foragers who collect only chanterelles, King Boletes, morels, and Oregon White Truffles. They may not know or care about other species. If we limited ourselves in this way, this would be a very short book. This raises the question of why we bother to include photographs and descriptions of species not suitable for table use. Here are a few good reasons:

- When you can identify fungi, a walk in the woods becomes a time of discovery and excitement. You develop "mushroom eyes," and a whole new world opens to you. Paul Stamets describes mushroom hunters as having burned mushroom images on their retinas, overlaying those images on the landscape, and when this happens, "the mushrooms seem to *jump* out at you" (A. Isaacson, "Return of the fungi" [*Mother Jones,* November–December 2009, p. 70]).
- You learn more about your environment. An interest in mushrooms leads to identification of tree species, because some mushroom species grow in association with particular types of trees. For example, certain chanterelles grow under oaks and some boletes are associated with pines. You also become sensitive to temperature and rainfall patterns. If you become interested in postburn mushrooms like morels, you start keeping track of fires on national forest land.
- Mushrooming as a hobby converts rain into a positive. In arid areas of the West, mushroomers eagerly await the rainy season.
- Getting into the woods, a habitat where nature is in control, lifts the soul. You will want to protect forest habitat against encroachment and ensure that future generations have access to natural places.
- You may make a discovery. In the western United States, a great number of mushroom species have yet to be prop-

erly identified. For example, most of the western species in the genera *Russula* and *Cortinarius* likely are unique and need to be named. Amateurs can explore new places and frequently discover new species (although it helps to talk to a professional to confirm that the find is a new species).

- Fungi are valuable to humans for reasons other than edibility. They can be used to make dyes, medicines, inks, crayons, and paper. Mycorrhizal fungi are essential to the survival of many tree species and therefore to the health of the entire forest. Other fungi cause serious plant, human, and insect disease. Many fungi are extremely beneficial to the planet because they are prime recycling organisms. You may find it satisfying to learn the names and characteristics of these diverse and interesting organisms.

- Species come in all sizes, shapes, and colors. They are visually interesting, and some are beautiful. We include a section on photographing and drawing mushrooms, making artistic spore prints, painting with mushroom spores, and using mushroom dyes.

Equipment

Mushrooming equipment is inexpensive—a basket, roll of wax paper or small paper bags for wrapping individual specimens (avoid plastic bags, which cause specimens to deteriorate quickly), a knife, paper for spore prints, and one or more field guides. Optional items include a journal for taking notes, camera, brush to clean dirt from specimens, walking stick for penetrating underbrush, and rake for finding truffles. Good raingear is essential—waterproof hat, jacket, rain pants, and boots.

A good field guide is a must. Many are available, and a single field guide may not be sufficient. Authors vary in their coverage of region and species. The number of mushroom species is so vast, and local and regional variation in field characteristics is so great—including differences in size and color—that no guide can cover them all. As in other natural history books, mushroom guides emphasize showy, distinctive fungi and ignore little brown (or white or gray) mushrooms. Most mushroomers have a favorite field guide but do not restrict themselves to it. A library of mushroom books costs less than a single foray to a distant location. Collect them all, or at least purchase a national, a regional, and a local guide if available.

Mushroom Eyes

The ability to see mushrooms in the field or forest requires mental preparation; you do not need 20/20 vision. People who seek tree mushrooms look up or across the trail, while those who seek ground-dwelling mushrooms inspect the forest floor. Others may search for humps in the duff (shrumps) or rake the earth around Douglas-fir for truffles. You develop mental templates of shape, color, and location for a desired species. For chanterelles, you "go for the gold"; for King Boletes (*Boletus edulis*), search for large, rounded humps in pine needles; for Black Trumpets (*Craterellus cornucopioides*), look for small dark shadows in places where there shouldn't be shadows; and for Matsutake (*Tricholoma magnivelare*), look for white circular caps barely breaking the duff.

When you find the first of a desired species, look around, because more are probably nearby—finding one means that

Mushrooms are sometimes discovered by finding a "shrump," or mushroom hump of elevated debris.

fruiting conditions are suitable for the species. Going on a foray is advantageous in this regard. When a person shouts "Eureka!" it means that conditions are good for finding more and explains the statistically improbable distributions in mushroom collecting—none or many. Don't go into the woods expecting huge fruitings that fill bushel baskets—it's great if you find them but discouraging if you don't. Instead, look for the first of a species that will betray the location of its fellows.

Making a Spore Print

Mushroom spores vary in color, and spore color doesn't necessarily match cap color. A white-capped mushroom can have brown spores, and a brown-capped mushroom can have white spores. Gill color isn't a reliable guide either, because it can change over time. The gills of Meadow Mushrooms (*Agaricus campestris*) are initially pink and then turn chocolate brown. When mushrooms cluster with one beneath another, you may find a spore deposit on the lower mushrooms. Otherwise, you should make a spore print to confirm spore color. This can be started in the field or done after you arrive home.

Spore prints are essential for both the beginner and the experienced mycophile (fungus lover) faced with an unfamiliar species. Obtaining a spore print is easy. The only equipment needed is a mushroom and two small pieces of paper, one black and one white. If the mushroom is large, cut off a section of the cap and place it over the two pieces of paper, with gills or pores facing down. With a medium-size mushroom, remove the cap from the stalk and place the entire cap over both pieces of paper. With small mushrooms, you'll probably need prints from two separate caps, one on white paper and the other on black paper. Covering the cap with a bowl maintains high humidity and prevents air currents from carrying away the spores before they fall on the paper.

Let the mushroom drop spores until a sufficient amount collects on the paper. Sometimes it takes less than 30 minutes; other times it may take hours or occasionally days. If the weather has been very dry, the mushroom may never drop spores, so

Making a spore print simply requires a mushroom cap, a piece of paper, and a little time. A bowl placed over the cap contains the spores by blocking air currents.

bring home several specimens—if one yields no spores, try another. You may also have to let wet or soggy mushrooms dry out, gills facing up, before you can obtain a spore print.

Knowing the spore color narrows the number of genera you need to consult to identify the mushroom—but there are a few caveats. Words for color can be confusing and overlap. For example, spores can be white, off-white, buff, or creamy. *Brown* can be light brown, cigar brown, reddish brown, or dark brown. As you gain experience, distinguishing among color names becomes easier. Even if you aren't a cigar aficionado, you can assume that cigar brown is darker than light brown but lighter than dark brown and less ruddy than reddish brown.

When a mushroom has been standing for a while, it may have already dropped most of its spores. You may be able to coax additional spores by leaving it under a bowl for a few days, but sometimes nothing works and you are left without a spore print. This won't be a problem for professional mycologists able to do DNA sequencing, but amateurs may find this frustrating. Don't try to guess spore color—move on to mushrooms whose spore color is known.

When Identification Fails

If a mushroom is unfamiliar and cannot be found in a guide book in your possession, follow these steps:

- Photograph the mushroom in its habitat; take one picture of the cap and another of the stalk and the underside.
- Collect several specimens, including examples of immature and mature fungi. Dig up enough of the base to make sure you capture any cup that is hidden at the bottom of the stalk, but don't dig so deep as to damage the mycelium. Make sure the specimens are typical and in good shape. Collecting poor, damaged, or old specimens will make identification more difficult.
- Record in a journal details about the mushroom, including its location and any trees nearby, growth characteristics (clustered or isolated, on wood or on the ground, etc.), any bruising or staining, feel of the cap (e.g., dry, slimy, smooth, scaly), and distinctive odor.
- At home, make a sketch that will sensitize you to details you may have missed. If possible, examine the spores, gills, cap, and stalk microscopically and sketch the details. Then use field guide keys to identify the mushroom. A common mistake is to let the key or guide influence your observations. Making your observations first minimizes the desire to find or manufacture characteristics mentioned in the field guide.

Spore color and a few key characteristics can bring you to a genus, but not necessarily a species name. This is especially likely in huge genera such as *Cortinarius* or *Inocybe*, which may not be of interest to collectors of edible mushrooms. As you become more knowledgeable and acquire additional resources and contacts, check back in your journal and try to identify mystery fungi missed earlier. Bring your journal to club meetings and to fungus fairs, where you can consult the person staffing the identification table.

Even the most experienced mycophiles occasionally find unidentifiable specimens. If you are seriously interested in a particular mushroom that you cannot identify from observable characteristics, and you do not have access to laboratory equipment, take it to someone who can examine the spores or other mushroom features microscopically or, even better, who can sequence the DNA and access databases such as GenBank, which has thousands of fungal species.

Trail Manners

No one likes to see fungi scattered in pieces along the trail. Many find it discouraging to see discarded mushrooms strewn about, obvious holes where mushrooms were dug up by humans, and tree trunks showing evidence that polypores were cut away. Pick-and-drop behavior produces litter that is upsetting to all nature lovers, as well as to other mushroomers. The practice is disrespectful of our environment, and park visitors will complain to the authorities, who may deny access.

COMMONSENSE TRAIL RULES

- Don't vacuum the forest. If your group has beginners, encourage restraint. One or two examples of a species among a group of people will be sufficient for identification.
- Leave immature specimens such as tiny morels or chanterelles time to mature and drop their spores. Follow the Rule of Thumb—if it isn't the size of your thumb, don't pick it.
- Leave attractive mushrooms fruiting alongside the trail so that others may enjoy them. Search for mushrooms hidden from public view.
- Replace your divots. When you dig up a mushroom, cover the hole with duff.
- Cover cut stalks with pine needles or dirt so that they are not visible from the trail.
- When raking around trees for truffles, try to be as gentle as possible. Rake from the tree outward and carefully replace the duff you have disturbed—

besides improving aesthetics, it will prevent plant roots and fungal mycelium from premature drying.

- Pick up mushrooms heedlessly discarded by others. Place them so they cannot be seen from the trail.
- Don't pry off chunks of bark to remove a polypore. It damages the mycelium and the tree.
- When a location has many specimens of a fruiting species, leave a few to drop their spores.
- Discard your unwanted mushrooms where others won't see them.

Eating Wild Mushrooms

Other than buying mushrooms in the supermarket, we have no foolproof test of mushroom edibility. Growing on trees, the presence of pores instead of gills, the absence of a volva, a cap that peels easily, pink gills, not tarnishing a silver spoon—all these folk rules have their exceptions, and sometimes dangerous ones. It is crucial to properly identify a mushroom before you consider it for table use. Never rely strictly on resemblance to a color photograph. Consider mushroom identification like recognizing a friend. You don't rely on general characteristics (5 ft. 8 in., dark complexion, brown eyes, etc.), because many people fit this description. You should use a combination of many nuanced characteristics and even then leave room for those special features that make your friend a unique individual.

Don't serve wild mushrooms to people who aren't familiar with them. Some may have hysterical responses even when the species is safe. Worry and imagined illnesses are not pleasant for those prone to experience them, and the welfare of your guests should be foremost in your mind. Some people are allergic to mushrooms, including popular species such as morels, which are generally considered safe. Expert opinion on edibility can change; at one time the chicken-of-the-woods or sulfur shelf (*Laetiporus* species) was considered a foolproof edible in terms of easy identification. Today, along the West Coast, questions have been raised regarding its edibility, especially when growing on eucalyptus. A similar fate has befallen the distinctive yellow-gilled Man on Horseback (*Tricholoma equestre*).

The best way to start collecting for the table is in the company of knowledgeable people on a foray organized by a mushroom club. Even then, do not take the word of others; be sure to double-check the field characteristics of anything you are going to eat, and consume only a small portion the first time. Always cook mushrooms rather than eating them raw. Be especially careful about what you bring home for the table when collecting in an unfamiliar location. And be aware of the telephone number of the nearest poison control center. Experienced collectors serve as consultants to these centers, identifying species to enable medical authorities to make a proper diagnosis and treatment plan.

Many forest mushrooms are bland and tasteless. They won't hurt you, but they aren't worth eating. A few are toxic and will make you sick, and a small number are lethal. If all mushrooms were tasteless or toxic, edibility wouldn't be an issue. Happily, some taste good, are good for you, and are easier than others to recognize. Repeating the warning that we have no foolproof test of edibility, we describe the characteristics of some more recognizable edible species along with look-alikes to be avoided. We exclude from our list those that require elaborate processing to remove known toxins.

EASY EDIBLES

Blewit
Clitocybe nuda

KEY CHARACTERISTICS: Purple color of all parts of the fruiting body, pink spores, absence of a ring on the stalk, and smooth, oily cap.

DANGEROUS LOOK-ALIKES: Some blue or lilac species of *Cortinarius*, which have rusty brown spores and a cobweblike partial veil and ring.

COMMENT: Distinctive fresh flavor; dries easily and stores well.

Boletes (see also King Bolete)

Pored, terrestrial genera, including *Boletus, Leccinum,* and *Suillus.*

KEY CHARACTERISTICS: A spongy layer of pores rather than gills or teeth.

DANGEROUS LOOK-ALIKES: Avoid boletes with red pores (e.g., *Boletus eastwoodiae*). Boletes can be parasitized by the mold *Hypomyces chrysospermus,* turning the stout caps into a white or yellow powdery mass, rendering the moldy mushroom inedible.

COMMENT: The spongy pore layer is often removed before drying or cooking fresh.

Candy Cap

Lactarius rubidus

KEY CHARACTERISTICS: Cinnamon brown azonate cap, watery latex that does not change color, and maple syrup-like odor that may be faint in fresh specimens but intensifies as the fruiting body dries.

DANGEROUS LOOK-ALIKES: Any *Lactarius* with yellow latex or white latex that turns yellow.

COMMENT: Most easily recognized by its maple syrup odor; can be used fresh or dried; good in desserts.

Cauliflower Mushroom

Sparassis radicata

KEY CHARACTERISTICS: Cream-colored, wavy ribbonlike branches (like egg noodles) growing at the base of tree trunks.

DANGEROUS LOOK-ALIKES: None.

COMMENT: Stores well in the refrigerator.

Chanterelles

Cantharellus species

KEY CHARACTERISTICS: Mostly yellow-orange in color, although there are white species; foldlike ridges rather than bladelike gills on the undersurface of the vase-shaped cap.

DANGEROUS LOOK-ALIKES: The poisonous Jack-O-Lantern (*Omphalotus olivascens*) is olive toned and clustered, has decurrent gills (not ridges like chanterelles), and grows on wood. The False Chanterelle (*Hygrophoropsis aurantiaca*) has forked gills instead of ridges and thin flesh and is orange. The Scaly Chanterelle (*Turbinellus floccosus*) has a thick scaly cap.

COMMENT: Choice edibles that are often meaty and firm; some have a pleasant apricot odor.

Chicken-of-the-Woods, Sulfur Shelf
Laetiporus conifericola, L. gilbertsonii

KEY CHARACTERISTICS: Polypore with bright yellow and orange coloration of both the top and underside of the cap; fruiting in clusters on wood.

DANGEROUS LOOK-ALIKES: Be wary of fruiting bodies growing on eucalyptus or hemlock. Try a small portion and monitor your reaction.

COMMENT: When cooked, this mushroom has the texture of chicken. Cut away tough and leathery caps and tough bases; use only young and tender margins of caps.

Deer Mushroom
Pluteus cervinus

KEY CHARACTERISTICS: Pinkish spores, brown cap, free gills, mild radishlike odor, and growth on wood.

DANGEROUS LOOK-ALIKES: *Entoloma* species, which grow on the ground.

COMMENT: Good edible.

Fairy Ring Mushroom
Marasmius oreades

KEY CHARACTERISTICS: Buff or tan cap, tough stalk, white spores, widely spaced gills that do not run down the stalk, and the ability to rehydrate after drying; fruiting in lawns, often in rings.

DANGEROUS LOOK-ALIKES: Other species can fruit in a circular pattern on lawns, so make sure of your identification. *Clitocybe dealbata* has close rather than widely spaced gills, lacks a tough stalk, and does not revive in water.

COMMENT: Very good when served in soups, casseroles, and stews.

Garlic Mushroom
Marasmius copelandii

KEY CHARACTERISTICS: Strong garlic odor, striate pale brown cap, white spores, and minutely hairy, brownish, tough stalk without a ring; fruiting on the ground in leaf litter.

DANGEROUS LOOK-ALIKES: *Galerina marginata* grows on wood and has rusty brown spores and a ring on the stalk.

COMMENT: Dries well and rehydrates in water; good as a flavoring.

Hedgehog Mushroom
Hydnum repandum

KEY CHARACTERISTICS: Creamy to dull orange color, spines rather than gills or pores on the undersurface of the cap; fruiting on the ground.

DANGEROUS LOOK-ALIKES: None.

COMMENT: Edible and choice. Be sure the mushroom has spines under the cap rather than gills.

Honey Mushroom
Armillaria mellea

KEY CHARACTERISTICS: Light brown to gold cap with circles of dark fibrils; tough, fibrous, ringed stalk; fruiting in dense clusters on wood.

DANGEROUS LOOK-ALIKES: *Galerina marginata* is smaller, fruits in troops rather than large clusters, and has rusty brown spores; the poisonous Jack-O-Lantern (*Omphalotus olivascens*) lacks a ring, has an olive cast, and has yellow, orange, or olive gills.

COMMENT: The stalks are tough and usually not eaten.

Horn of Plenty, Black Trumpet
Craterellus cornucopioides

KEY CHARACTERISTICS: Blackish tubular fruiting bodies with smooth or faintly wrinkled undersurface. An individual fruiting body is thin and light, but because it commonly occurs in clusters, quantities can be collected.

DANGEROUS LOOK-ALIKES: None.

COMMENT: Tastiest when dried and reconstituted. Good in powdered form.

King Bolete, Porcini, Cep
Boletus edulis

KEY CHARACTERISTICS: Reddish brown cap, reticulate stalk, and whitish to olive-green pores that do not blue when bruised; the stalk is often bulbous when young.

DANGEROUS LOOK-ALIKES: Avoid any bolete with red pores.

COMMENT: One of the best edibles but often buggy.

Lion's Mane
Hericium erinaceus

KEY CHARACTERISTICS: "Icicles" or soft spines on a white cushion growing on wood.

DANGEROUS LOOK-ALIKES: None.

COMMENT: Young specimens are best for the table. Also cultivated and available in markets.

Lobster Mushroom
Hypomyces lactifluorum

KEY CHARACTERISTICS: A bright orange to red mold that roughly resembles the shape of the host mushrooms that it parasitizes.

DANGEROUS LOOK-ALIKES: Avoid white, pink, and yellow molds.

COMMENT: This mold most often parasitizes *Lactarius* and *Russula* species, but since you cannot be certain of the host's identity, be careful. Most important, make sure that the fungus has completely replaced its host. The interior must be firm without any trace of a soft or watery consistency.

Matsutake
Tricholoma magnivelare

KEY CHARACTERISTICS: White cap and stalk usually spotted with reddish brown stains, presence of a ring, and a spicy odor.

DANGEROUS LOOK-ALIKES: White amanitas; make sure there is no volva at the base of the stalk.

COMMENT: Chewy but very flavorful in soups and stews.

Morel
Morchella species

KEY CHARACTERISTICS: Pyramid-shaped cap with deep concave pits bounded on all sides with intersecting ridges, attached for its length to the stalk; texture firm-fleshy to rubbery.

DANGEROUS LOOK-ALIKES: *Gyromitra* species that are brainlike, with folds and wrinkles, in contrast to the pitted conical caps of morels.

COMMENT: Stores well when dried and reconstitutes in water; discard water before cooking. Excellent with egg and fish dishes.

Oyster Mushroom
Pleurotus ostreatus

KEY CHARACTERISTICS: Growing shelflike on wood from a short lateral stalk; typically fan shaped, white, gray, tan, or brown, surface smooth, spores white.

DANGEROUS LOOK-ALIKES: *Lentinellus* species have jagged-edged gills and a finely hairy cap surface.

COMMENT: Delicious edible. Also available fresh in supermarkets or grown in home kits on inoculated bags of straw or sawdust.

Pine Spike
Chroogomphus vinicolor

KEY CHARACTERISTICS: Reddish, orange, or reddish orange color of all parts of the fruiting body, decurrent gills, black spores, and very tough brownish orange stalk; fruiting under conifers.

DANGEROUS LOOK-ALIKES: Some *Cortinarius* species may superficially resemble the Pine Spike but they have rusty brown spores.

COMMENT: Dries well; good in soups and stews.

Puffballs
Calbovista subsculpta, Lycoperdon perlatum, Lycoperdon pyriforme, and others

KEY CHARACTERISTICS: Round or pear-shaped fruiting body with a white spore mass that discolors in age.

DANGEROUS LOOK-ALIKES: Immature *Amanita* species (eggs). When you cut a puffball in half lengthwise, the skin should be thin and the spore mass a homogeneous white, with no trace of an embryonic cap or stalk, as in *Amanita* species. Cutting specimens in half will also avoid confusion with inedible *Scleroderma* species, which have a tough skin (peridium) enclosing a purple-black spore mass.

COMMENT: Eat only fresh specimens with a homogeneous pure white interior.

Shaggy Mane
Coprinus comatus

KEY CHARACTERISTICS: Cylindrical to oval whitish cap covered with brownish flat or recurved scales, deliquescing and blackening from margin to the center.

DANGEROUS LOOK-ALIKES: The Inky Cap (*Coprinopsis atramentaria*) is toxic when consumed with alcohol. Unlike the Shaggy Mane's white cap with scales, the Inky Cap has a shiny gray cap with or without a brownish tint.

COMMENT: Eat only fresh white specimens; cook quickly because the mushroom will turn to black ink (autodigest) with age.

Corn Smut

Upscale restaurants call it corn fungus, corn truffle, even corn caviar; farmers refer to it as corn smut, and mycophiles know it as *Ustilago maydis*. A fungal infestation, it appears as shiny blue-gray boils or galls on corn ears. At maturity the galls burst, spreading blue-black spores that give the corn ear a burned appearance. *Ustilago* comes from the Latin *ustilare,* to singe or scorch. Considered a pest by farmers in the United States, it is a delicacy among native peoples in the Southwest and in Mexico. In response to a growing gourmet market, several farmers now grow it commercially under the names *huitlacoche* or *cuitlacoche.* You can buy it canned or occasionally fresh at specialty produce shops, or you can ask a grower at a farmers' market to save smutty ears for you.

Corn smut caused by the plant pathogenic fungus *Ustilago maydis*.

Kitchen Work

When collecting mushrooms, you can take steps to make preparation and storage easier. Cut away the dirt-covered stalk and brush off caps before putting them in your basket so that dirt and forest debris do not contaminate clean mushrooms. Avoid placing obviously buggy specimens in your basket because the infestation may spread. When you get home, go outside and shake your mushrooms in a wire basket to remove remaining forest debris. If some dirt remains that cannot be removed with a brush or damp paper towel, you can rinse mushrooms in water, but cook them immediately after washing, or the wet mushrooms will spoil. More delicate edibles such as Deer Mushrooms (*Pluteus cervinus*) and Oyster Mushrooms (*Pleurotus ostreatus*) become soggy in water, so use a mushroom brush to remove dirt. Inspect your mushrooms for pinholes and other signs of a major bug invasion. Most mycophagists (people who eat mushrooms) tolerate a few pinholes, appreciating the extra protein, but will cut away pieces or discard a mushroom showing signs of a major infestation.

Always cook foraged mushrooms before you eat them. The website of the Colorado Mycological Society provides additional advice on collecting for the table:

- Stay away from LBMs ("little brown mushrooms"); these include many poisonous species that are difficult to distinguish from edible species.
- Carefully examine each mushroom you pick. Remember, poisonous and edible mushrooms often grow side by side.
- Don't eat any mushroom that is past its prime. Eating edible species that have spoiled causes many mushroom poisonings.
- Look for specific recognizable species. When you go out picking berries, you collect strawberries or blackberries. You don't pick every kind of berry you come across, mix them in a basket, and expect some expert to tell you which berries are edible. Follow the same logic with mushrooms.
- Use caution and common sense. If in doubt, throw it out!

Drying Mushrooms

This is the simplest method of preservation, and for morels and King Boletes it intensifies the flavor. You can use the entire mushroom, but for mushrooms that have tough inedible stalks, such as the Fairy Ring Mushroom (*Marasmius oreades*), eat only the cap and use the rest to make broth. With clustered edibles such as the Honey Mushroom (*Armillaria mellea*), cut away exterior leathery caps and dry only tender interior ones. Small mushrooms such as Garlic Mushrooms (*Marasmius copelandii*) can be dried whole. Cut large mushrooms lengthwise into quarter-inch slices and place on a drying rack. You can buy a multiple-shelf vegetable dehydrator or construct your own by securing light bulbs at the bottom of a wooden cabinet with wire shelving so that the hot air rises through the mushrooms and exits at the top. You can also dry mushrooms placed a few inches over a radiator or otherwise warm and dry area in your house. Store *fully dried* mushrooms in labeled airtight plastic bags or jars away from light. You can also grind dried mushrooms in a

Boletes drying on a screen behind a sunny window.

blender to make mushroom powder as a condiment, or sun-dry thin slices of firm mushrooms to eat as "mushroom chips."

When drying, be prepared for a dramatic shrinkage in volume. Because some mushrooms are more than 95 percent water, a dried mushroom may be one-twentieth its former weight. This makes storage easier—a large quantity of dried mushrooms will fit into a small jar, but they never regain their original size and weight when revived. Exceptions to this are species of *Marasmius,* which reconstitute in water.

Freezing Mushrooms

Some mushroom species cannot be dried (e.g., Shaggy Manes [*Coprinus comatus*] will turn to ink) or, like chanterelles and Hedgehog Mushrooms (*Hydnum repandum*), lose much of their flavor when dried. Oyster Mushrooms are not a good drying candidate, either. These species are best sautéed in butter (with or without white wine and parsley), bagged, and stored in the freezer. Be sure to precook them first, which preserves their flavor and kills any bugs hiding in the cap. Another possibility is to blanch mushrooms before freezing. Drop them into boiling water with a teaspoon of lemon juice and cook for 3 minutes. The liquid can later be used as a broth.

Store fresh mushrooms in paper bags; avoid airtight containers or plastic wrap, which may promote mold. Chanterelles and Black Trumpets store well in the refrigerator. Cauliflower Mushrooms (*Sparassis radicata*) can keep for weeks, but most mushrooms must be used within a few days.

Canning and cooking mushrooms are both huge topics with a surfeit of excellent books and websites providing detailed instructions and recipes. Some of our favorite cookbooks:

Bessette, A. R., & Bessette, A. E. (1993). *Taming the Wild Mushroom.* Austin: University of Texas Press.

Czarnecki, J. (1995). *A Cook's Book of Mushrooms.* New York: Artisan.

Farges, A. (2000). *Mushroom Cookery.* New York: Workman.

Freedman, L. (1987). *Wild about Mushrooms.* Berkeley, CA: Aris Books.

McLaughlin, M. (1994). *The Mushroom Cookbook.* San Francisco: Chronicle Books.

Nims, C. C. (2004). *Wild Mushrooms.* Mount Claremont, WA: Westwinds Press.

Hallucinogenic Mushrooms

The psychoactive effects of a few mushroom species have long been recognized. Small sculptures celebrating sacred mushrooms found in the highlands of Guatemala are more than 3000 years old. When the Spanish conquistadors reached the New World in the sixteenth century, they described indigenous peoples using mushrooms in religious ceremonies. Dominican Friar Diego Durán in his *History of the Indies of New Spain* wrote, "By dint of these mushrooms, they saw visions and the future was revealed unto them." The Spanish priest Bernardino de Sahagún described in his *Florentine Codex* the role of sacred mushrooms in Aztec culture: "The first thing to be eaten at the feast were small black mushrooms they called *nanactl* [literally God's flesh; the species were probably *Psilocybe caerulescens* and *P. mexicana*—Ed.] that bring on drunkenness, hallucinations and even lechery. . . . When the drunkenness of the mushrooms had passed, they spoke with one another about the visions they had seen."

The major hallucinogenic mushrooms in the western United States in terms of frequency of use are several *Psilocybe* species and *Amanita muscaria* (which is toxic as well as hallucinogenic).

LSD (lysergic acid diethylamide) was first synthesized in 1938 by the Swiss chemist Albert Hofmann in a study on the ergot fungus (*Claviceps purpurea*) that grows on rye and other plants. Although ergot has been responsible for many mass poisonings, in very small doses it has useful medical properties, and the pharmaceutical industry was interested in exploiting its benefits. In a laboratory accident, Hofmann absorbed a small quantity of LSD through his fingers, and he later discussed its potent hallucinogenic properties. In 1957, Gordon Wasson, a banker, publicized the use of hallucinogenic mushrooms after a trip to Mexico, where he witnessed the consumption of *Psilocybe* mushrooms during a healing ceremony. Hofmann subsequently isolated psilocybin from *Psilocybe,* creating more interest in fungi that could induce hallucinations. Street names for *Psilocybe* mushrooms include shrooms, magic mushrooms, cubes (*P. cubensis*), and liberty caps (*P. semilanceata*). They produce emotional as well as perceptual changes that may be very unpleasant. Effects tend to last a few hours. Possession of *Psilocybe* mushrooms or psilocybin is currently illegal in the United States.

Use of *Amanita muscaria* to change consciousness is most often cited in Scandinavia, Russia, and Asia. A 1730 book by Swedish Colonel von Strahlenberg, who spent twelve years in Siberia as a prisoner, contained a description of native people drinking tea made from *A. muscaria* and also drinking the urine of those who had consumed the mushroom to minimize some of the unpleasant side effects of ingesting the actual mushroom. "They lay up large provisions of these mushrooms for the winter," von Strahlenberg wrote, "When they make feast, they . . . boil these mushrooms and drink the liquor which intoxicates them." The eighteenth-century Danish historian Samuel Ödman attributed the ferociousness of Viking Berserkers to eating *A. muscaria* before battle. The species grows throughout the western states. Its recreational use is limited because of its unpleasant side effects.

MUSHROOM CULTIVATION

A number of edible mushrooms can be grown at home. Saprobic fungi, those that grow on dead organic matter such as compost or decaying wood, are suitable candidates for home cultivation. Examples include Shiitake (*Lentinula edodes*), several species of oyster mushrooms (*Pleurotus*), and Reishi (*Ganoderma lucidum*), among others. Fungi that are mycorrhizal, that is, living in a mutually beneficial partnership between the fungus and a plant, have not been commercially cultivated. Thus, some mushrooms, such as chanterelles, King Bolete (*Boletus edulis*), and Matsutake (*Tricholoma magnivelare*), are collected in the forest, and no one has unlocked the mystery of their fruiting requirements.

The Button Mushroom, *Agaricus bisporus*, is cultivated commercially on carefully formulated compost. Although horse manure often comes to mind as the substrate (food source) used on a mushroom farm, most Button Mushrooms purchased in a store were not grown on a substrate that included horse manure—there simply isn't enough horse manure to go around. When it is used, it is a very small fraction of the compost, which primarily consists of wheat straw amended with agricultural waste products, such as grape pomace, feather meal, cocoa bean hulls, and other inexpensive organic sources. The straw and amendments are soaked with water and allowed to ferment for about three weeks. Because the pile has to be fairly large to generate the heat needed for certain chemical reactions to occur and has to be turned inside out and upside down on a regular basis, the labor required for good compost production becomes intense. Compost piles in home gardens usually do not generate sufficiently high temperatures or possess the proper carbon:nitrogen ratio needed for consistent mushroom cultivation. Commercially, heavy equipment is used to flip the compost pile. Other mushrooms that can be grown on compost include the Shaggy Mane (*Coprinus comatus*), which is easily cultivated but has a limited shelf life since it auto-digests at maturity.

Better candidates for home cultivation are those that can be cultivated on wood, wood products, or other kinds of cellulose. Oyster mushrooms (various species of *Pleurotus*) are the easiest mushrooms to grow at home. They grow on many readily available sources of cellulose, such as paper, sawdust, coffee grounds, and corn cobs. The substrate is typically chipped or shredded and pasteurized in hot water or sterilized in a pressure cooker. To pasteurize the substrate, soak it in water heated to 160°F for 1 to 2 hours. This eliminates many of the fungi and bacteria that would otherwise compete with the species under cultivation. Coffee grounds are an ideal substrate because they have already been steeped in hot water (pasteurized), and most coffee houses are happy to give them away. Once pasteurized or sterilized, the substrate is allowed to cool. Spawn, or inoculum of the fungus, is added to the substrate as it is packed into plastic bags or other suitable containers. Air holes are necessary because all mushroom-producing fungi need oxygen to grow. Many cultivators use special bags that have a filter patch that excludes mold spores in the air but allows gas exchange.

Spawn added to the substrate is often grown on sterilized grain (usually rye or millet) produced by the cultivator or purchased from a mushroom supply house (see Resources). For many mushrooms, spawn is added at a rate of about 3 percent of the dry weight of the substrate. Depending on the species of mushroom, the substrate is usually fully captured by the myce-

Pink Oyster Mushrooms (*Pleurotus djamor*) cultivated on straw.

lium within 2 to 6 weeks, although some species take longer. This step, called spawn run, takes place at temperatures around 70 to 75°F. To initiate mushroom formation, temperatures are often dropped 5 to 10°F, carbon dioxide levels are lowered by introducing fresh air, and light is provided. It is a misconception that mushrooms grow in the dark. Almost all mushrooms require light for proper fruiting body development. An exception is the Button Mushroom, which can grow in total darkness with no ill effects. During fruiting, mushrooms require high humidity. This is achieved by a commercial humidifier or wetting the floor of the mushroom-growing room and allowing natural evaporation to humidify the air. In the dry climate of the western United States, providing a humid environment is often the limiting factor for successful home mushroom cultivation.

While oyster mushrooms are easily grown on pasteurized straw and other sources of cellulose, most other specialty mushrooms are grown on sawdust. Shiitake, for example, is grown in plastic bags containing a sterilized mixture of sawdust, millet, wheat bran, and water. The sawdust can be made from many types of wood, but oak and alder are the most popular. Cedar and redwood are not suitable because they are resistant to fungal decay to various degrees. Pine wood is also unsuitable since most cultivated fungi are inhibited by the resins in the wood. The coarseness of the sawdust should be about that from a chain saw. Larger chips are sometimes included. A common recipe for a sawdust substrate, by dry weight, is 76 percent sawdust, 12 percent millet, and 12 percent bran, with a final moisture content of 65 percent. Other mushrooms that grow and fruit on this substrate include Lion's Mane (*Hericium erinaceus*), Enoki (*Flammulina velutipes*), Nameko (*Pholiota nameko*), and Reishi. Enoki is purposefully grown in the dark at very cold temperatures, resulting in a completely white, small-capped, long-stalked fruiting body. Note that the wild type (p. 140), which is a fairly common mushroom in the western United States, is a rich brownish orange with a dark brown stalk.

The quickest and most convenient way to start mushroom cultivation is to purchase bags that contain a substrate already colonized by the mushroom of your choice. All that is required is a cool and humid environment and perhaps a little added moisture. These kits come with instructions and are readily available from several mushroom supply houses.

TOXINS

Mushroom poisonings are caused by eating fruiting bodies of certain species of fungi, popularly called toadstools. The word *toadstool* has two versions of its origin. One involves *toad + stool*, since some toads are poisonous, and the other comes from the German *tod*, meaning death, + *stool*. We have no quick and easy ways to distinguish between edible and poisonous species. When collecting for the table, you must be certain of what is in your basket. Unfortunately, poisonings occur every year, often due to misidentification. Sometimes a collection is off-type because of natural variation or because it fruited in full sun or deep shade; other times immigrants collect toxic mushrooms that appear similar to edibles in their native land, with disastrous consequences. Eating over-the-hill mushrooms may also result in stomach upsets.

Mushroom toxins are categorized by their effects on the body. The four categories discussed here are (1) the toxins that cause cell death and organ failure, (2) those that affect the nervous system, (3) those that cause gastrointestinal upsets, and (4) the disulfiram-like toxins, which cause no symptoms unless alcohol is consumed with or after eating the mushrooms.

The toxins responsible for most fatal poisonings are the amatoxins, a group of toxins that include the amanitins. These are found in the Death Cap (*Amanita phalloides*) and Destroying Angel (*Amanita ocreata*), the species responsible for most mushroom fatalities in the West, and several others. Amanitins inhibit protein synthesis, resulting in cell death and eventually organ failure, especially liver and kidney failure. Symptoms of this kind of poisoning are not immediate, which makes treatment difficult because the victim does not make the connection between eating mushrooms and the onset of symptoms. After this latent period (6 to 15 hours or more), the victim may experience diarrhea, vomiting, and abdominal pains. Symptoms then subside and the victim appears to recover for a short time, but during this time the victim's organs are deteriorating. Liver and kidney function may fail, causing a life-threatening situation. Other mushrooms with high levels of amanitins include *Galerina marginata*, a small wood inhabitant usually too small in size to be considered for the table, and some species of *Lepiota*, *Conocybe*, and possibly others. Hospitals rely on knowledgeable mycologists to identify the toxic mushroom, but when only bits of the specimen remain or stomach contents offer hints of the source of the toxin, laboratory methods are used. Minute concentrations of one of the amanitins found in *Amanita* species can by detected by mass spectrometry, and minute pieces of mushroom tissue can be identified by molecular (DNA) procedures.

Another group of toxins that cause general toxicity to cells is found in some false morels (e.g., *Gyromitra esculenta* and *G. infula*). One identified toxin in these mushrooms is gyromitrin, a volatile form of the chemical hydrazine related to rocket fuel. Ingestion of this toxin causes gastrointestinal distress preceded by a latent period, like the delay that accompanies poisoning by amanitins. The liver and the nervous system are affected. Although the toxin is extremely heat unstable and will volatize as

Gyromitra infula, one of the false morels, contains a highly volatile toxin called gyromitrin. Although the toxin evaporates during cooking, this mushroom should be avoided for the table.

soon as the mushroom hits a hot pan, some gyromitrin may remain; therefore, these mushrooms are not recommended for the table. A related compound, called agaritin, is present in small amounts in species of *Agaricus*, including the commercial Button Mushroom. This compound is a known carcinogen if eaten in massive quantities. It, too, is heat unstable, and much of it quickly volatizes when the mushrooms are cooked. This is one of the reasons why it is a good idea to cook all mushrooms.

Some species of *Cortinarius* contain a toxin called orellanine. This toxin has a long latent period of up to 14 days. It is characterized by an intense thirst and burning of the mouth, followed by nausea and loss of consciousness. Kidney damage or failure may follow. Because species of *Cortinarius* are so poorly known, especially in the southwestern United States, no species of *Cortinarius* should be considered for the table.

Other toxins cause poisoning of the nervous system. The toxin muscarine occurs in *Inocybe geophylla*, *Clitocybe dealbata*, and some other species. This toxin results in profuse sweating, salivation, and, if eaten in large doses, nausea, gastrointestinal pain, and difficulty in breathing. Deaths are rare. *Amanita muscaria* has insignificant amounts of muscarine, but it was the first toxin identified in that species, hence the name.

Ibotenic acid and muscimol are present in *Amanita muscaria*, *A. pantherina*, and possibly other species. Symptoms include drowsiness, dizziness, gastrointestinal upsets, and periods of excitability, hyperactivity, and hallucinations. Young children may experience convulsions and other neurological disorders. Ibotenic acid and related compounds are not destroyed by heat, but they are soluble in boiling water. Some claim that, by throwing out the cooking water, these mushrooms can be made edible, but because of the elaborate preparation involved, they are not recommended for the table.

Another toxin that affects the nervous system is the well-known psilocybin, found in the hallucination-inducing species of *Psilocybe*, *Panaeolus*, *Stropharia*, and others. In the body, psilocybin is converted to psilocin, which causes hallucinations and symptoms similar to alcohol intoxication. The bluing reaction when the flesh of *Psilocybe* is bruised indicates the presence of psilocybin. It should be noted that other mushrooms bruise blue for different, unknown reasons.

Many species of fungi can cause gastrointestinal upsets. In most cases the specific toxic compound is unknown. Species that cause nausea, diarrhea, and abdominal cramps include *Chlorophyllum molybdites*, *Tricholoma pardinum*, *Russula cremoricolor*, and *Agaricus xanthodermus*, among many more. The onset of the symptoms is typically soon after ingestion.

The last category of toxins includes the disulfiram-like toxin found in the Inky Cap (*Coprinopsis atramentaria*). Although this species is generally considered to be an edible, it produces a chemical called coprine, which is converted in the body into a compound that interferes with the breakdown of alcohol. If alcohol is consumed with *C. atramentaria* or even up to 2 days after a meal of the mushroom, headache and nausea can ensue.

Mushroom poisonings occur every year in the western states. When considering wild mushrooms for a meal, be safe rather than sorry. If you are a beginner, seek expert opinion before deciding that a species is suitable for the table. Don't experiment or gamble, as the consequences of error can be serious. There is a saying that there are old mushroomers and bold mushroomers, but there are no old bold mushroomers.

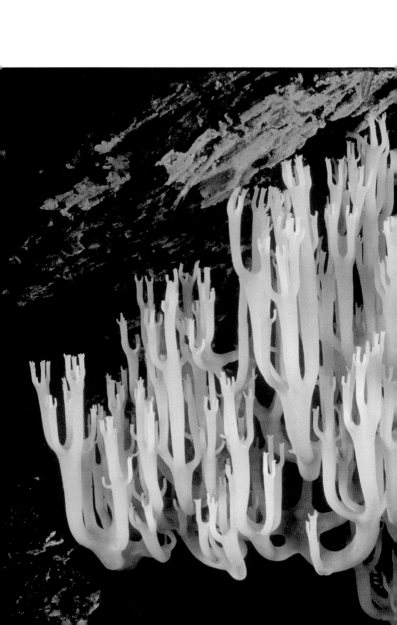

THE DEFINITION OF *species* is a highly debated subject. One commonly used definition is "an aggregate of interbreeding individuals," which in practice is difficult to determine in fungi. Today, the definition commonly includes a component that describes genetic isolation from other species. In addition, some mycologists insist that a definition of a fungus species include a description of a unique habitat or ecological role, which helps characterize the individuality of a species.

Like other mushroom guides, this one organizes species by observable characteristics, such as the presence of platelike gills, pores, and spines. The convenience of gross morphology facilitates identification, but the apparent relationships among many species based on physical characteristics alone are artificial. Fungi, like species of plants and animals, can be organized by natural relationships, something akin to our own personal genealogies. In our own family trees, an individual is connected to two ancestors (mother and father), who in turn are connected to each of their parents, and so on. In contrast, the true relationship between two species of fungi is based on their phylogeny, or evolutionary history. This differs from genealogies because in a phylogeny an individual species is connected to a single ancestor, and to one or more sister species. In a phylogenetic tree, the lineage of a species is unique with respect to lineages of other species. These lineages are often more interesting than traditional systems of classification where groups of species are organized by physical appearance. Species of *Agaricus,* for example, share a more recent common ancestor with species of *Lycoperdon,* a genus of puffballs, than they do with other gilled mushrooms such as *Amanita* and *Cortinarius* (see Figure 5). This unexpected relationship is a good example of how physical form often cannot be used for natural organization.

To understand mushroom phylogeny, and therefore natural relationships, we turn to genetic evidence from DNA sequences and modern systematics (the science of classification). Studies based on DNA data have supported some previous assumptions on fungal classification based on gross morphology but have totally invalidated others. In the phylogenetic tree shown here, common ancestry of selected basidiomycetes is indicated by hierarchical shared branches in the tree. Note that species of *Russula* do not share recent common ancestors with *Agaricus* and other gilled mushroom species, despite the fact that they out-

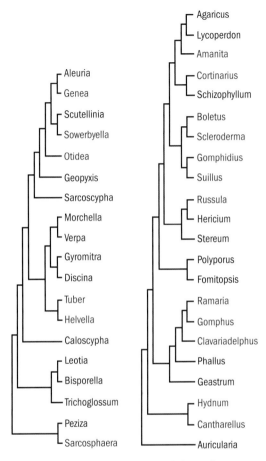

Figure 5. Phylogenetic tree depicting evolutionary lineages of some genera of ascomycetes (left) and basidiomycetes (right). Genera that share branches have a common ancestry; that is, closely related groups are located on branches close to one another. Genera in red are largely or entirely mycorrhizal; those in black are saprobic. The trees are based on sequences for ribosomal DNA. For more information, see Leho Tedersoo, Tom W. May, and Matthew E. Smith, "Ectomycorrhizal lifestyle in fungi: Global diversity, distribution, and evolution of phylogenetic lineages" (*Mycorrhiza* 20 [2010]: 217–263).

wardly appear very similar. Surprisingly, species of *Russula* share common ancestry with *Hericium,* a tooth fungus.

DNA is the molecule that contains the genetic code, a kind of blueprint, of all life. The genetic code makes every individual unique, except for identical twins, who share exact copies of their DNA. The DNA molecule consists of nucleotides containing a sugar component, a phosphate group, and one of four nitrogen-containing bases, adenine (A), thymine (T), cytosine (C), and guanine (G). What makes an organism unique is the order of the nucleotides. The millions of bases present in the DNA of all organisms provide an abundance of data for mycologists, who look for similarities as well as differences in DNA makeup. Shared portions of DNA sequences create groups, called clades, of related species.

The goal of DNA systematics is to discover the one true phylogeny of fungi, an ambitious and perhaps unattainable objective since it can't be proven in any case. One goal that is readily within our reach is the revelation of common ancestry among groups of species. For example, in the phylogenetic tree in Figure 5, one group connected by common ancestry consists of *Morchella, Verpa, Gyromitra, Discina, Tuber,* and *Helvella.* DNA evidence also provides insights into variation of individual species and refines our concepts of others. Recent studies have revealed that some species thought to be distinct from one another are actually different color variants of the same species, and some well-known species are actually complexes of many species. This is fairly common when species that seem to have a worldwide distribution are carefully examined. In many cases, populations isolated from one another by distance represent separate species, despite outward similarity. Whenever you see a species name, allow for the possibility that it represents a group of several related species lumped together for parsimony and clarity. *Amanita pantherina,* for example, is a complex of an unknown number of species in the western United States, and *Russula brevipes* is a complex of at least four species. Thus, many species of fungi are waiting to be named. It has long been recognized that many existing species names in North America were "borrowed" from European names and need to be changed. Consequently, the names in use for many of species in the western United States are in a state of flux as new evidence supports the argument that much of the fungal flora of the West is unique.

Based in part on DNA evidence, it is apparent that all but a handful of several hundred species of *Cortinarius* that occur in the West are probably unique and await proper names. One common theme in contemporary mycology is that many names in current use are incorrect, but we have none better. Until species in our part of the world gain more attention, we are often forced to use names that will get the reader in the ballpark, with the realization that some existing names are no more than placeholders.

One reason it is important to give a species a unique name is communication. We need to know exactly which species we are looking at to be able to share information with others. While mushrooming is an enjoyable pastime on its own, species should be conserved not only because they may hold untold virtues for the pharmaceutical world or are potential food sources, but also because they are an integral part of life on Earth. To conserve them, we need to know just what exists on the planet, and in the case of mushroom species, we have only scratched the surface.

QUICK IDENTIFICATION GUIDE TO MAJOR GROUPS

Gilled mushrooms. Familiar fungi with a cap and platelike gills where spores are produced. A vertical or lateral stalk is usually, but not always, present.

Chanterelles. Spores borne on ridges or folds on the underside of the cap and running down the stalk. The cap is often centrally depressed or funnel shaped.

Tooth fungi. Spores produced on downward-facing teeth, spines, or "icicles." Distinct cap present or absent.

Club and coral fungi. Fruiting bodies erect and club-shaped or fingerlike or profusely branched and ocean coral-like.

Jelly fungi. Rubbery, gelatinous, disklike, tonguelike, or shapeless blobs growing on wood, often colorful and translucent. Many shrivel when dry and rehydrate when wet.

Crust fungi. Thin, effused, exposed spore-bearing layer lying flat on wood substrate.

Boletes. Spores borne in pores on the spongy underside of a fleshy cap. The pore layer is separable from the cap.

Polypores and conks. Spores borne in pores on the underside of a tough or woody cap. The pore layer is inseparable from the cap. Some have a stalk, others are stalkless, and some lie flat on their substrate.

Bird's nest fungi. Small, nestlike fruiting body with spore-bearing "eggs."

Stinkhorns. Spore-bearing surface a layer of slime that smells like decaying flesh. The fruiting body is unbranched, upright, and clublike or ornately branched.

Puffballs and earthstars. Fruiting body a rounded sac containing spores, which are not ejected from the sac. The spores are released through an apical pore or when the wall of the sac wears away. In earthstars, the outer layers split to form rays.

Morels, cup fungi, and allies. Spores borne in microscopic sacs called asci. Fruiting bodies contain tiny spore-bearing flasks or the spores are produced in a layer in simple or complex cups with or without stalks.

GENERAL KEYS

These dichotomous keys consist of two or, rarely, more state-ments consisting of contrasting characteristics. With few excep-tions, you can see the features in the field without need for a hand lens. A couplet directs you to another couplet (numbered without parentheses) or, ultimately, to an identification (with page numbers in parentheses), which you should confirm by checking it against the illustration and description provided for each species. Keep in mind that this field guide is not compre-hensive, and the keys will not identify many specimens. You can identify some mushrooms from the information on similar spe-cies discussed in the text accompanying each illustrated species.

Spores produced on basidia (club-shaped, spore-bearing cells)...
.... Basidiomycetes (below)
Spores produced inside asci (spore-bearing sacs).......................
....Ascomycetes (61)
Spores produced in fruiting bodies from a slime body (plasmo-dium)..Slime Molds (408–411)

Basidiomycetes

Basidia and spores borne externally on exposed gills, spines, pores, etc.; spore print obtainable ....**Hymenomycetes** (below)
Basidia and spores borne internally (inside the fruiting body or inside a spore case); spore print not obtainable ....
....**Gasteroid fungi** (61)

Hymenomycetes

Platelike gills present............................**Gilled mushrooms** (below)
Platelike gills absent (but pores, spines, warts, folds, ridges, wrinkles, etc., may be present)...........**Nongilled mushrooms** (58)

Gilled Mushrooms

Spores white, yellow, green, or pale pink..............................below
Spores reddish (red, salmon, or pinkish brown)...................(56)

Gilled Mushrooms with White, Yellow, Green, or Pale Pink Spores

1a. Universal veil forming a volva and often leaving warts or patches on cap...................................see key to **Amanita spp.** (66)

1b. Volva absent..2

2a. Gills free; partial veil leaving a ring on stalk ....
....see key to **Lepiota spp. and allies** (83)

2b. Gills attached to the stalk; partial veil present or absent 3

3a. Fruiting body rigid and brittle (snaps clean like chalk); gills exuding a latex when injured or not 4

3b. Flesh usually fibrous or cartilaginous or thin and fragile....5

4a. Gills, flesh, or both exuding a latex when injured..................
....see key to **Lactarius spp.** (91)

4b. Flesh not exuding a latex.........see key to **Russula spp.** (100)

5a. Gills waxy and often widely spaced; caps often brightly colored.. see key to **Waxy caps** (119)

5b. Gills not waxy, close to widely spaced....................................6

6a. Stalk absent; cap often fan shaped...7

6b. Stalk present..8

7a. Gills split, cap hairy.............. **Schizophyllum commune** (131)

7b. Gills not split; cap smooth, fan or oyster shaped ....
....**Pleurotus ostreatus** (132)

8a. Ring present 9

8b. Ring absent.. 12

9a. Cap and stalk below ring covered with a layer of powdery granules.. **Cystoderma spp.** (133)

9b. Cap and stalk not covered with a layer of granules 10

10a. Growing on wood, often in large clusters; cap sparsely hairy ....**Armillaria mellea** (134)

10b. Growing on the ground... 11

11a. Cap yellowish; stalk shaggy with soft scales; odor mild.......
....**Floccularia spp.** (135)

11b. Not with all the above characteristics.....................................
....see key to **Tricholoma spp.** (159)

12a. Stalk fleshy or fibrous (like the texture of the cap), usually more than 5 mm wide.. 13

12b. Stalk cartilaginous or wirelike and tough or fragile and thin (often unlike the texture of the cap), usually less than 5 mm wide 24

13a. Fruiting body reddish tan, cinnamon, or purplish; gills well spaced; stalk tough, usually less than 1 cm wide, often twisted and always distinctly fibrous; growing on the ground...................
.... ***Laccaria* spp.** (136)

13b. Not with all the above characteristics............................... 14

14a. Fruiting body robust and dense; cap and stalk dry, dull, and unpolished; cap white or brown; white mycelial mat usually present at base of stalk; taste usually bitter or unpleasant.............
.... ***Leucopaxillus* spp.** (138)

14b. Not with all the above characteristics............................... 15

15a. Cap gray to brown, medium sized to robust; growing in dense clusters on the ground; gills and stalk whitish....................
.... ***Lyophyllum decastes*** (139)

15b. Not with all the above characteristics............................... 16

16a. Cap bright orange-brown, viscid; stalk dark brown and velvety; in clusters on wood ....***Flammulina velutipes*** (140)

16b. Not with all the above characteristics............................... 17

17a. Gills bright orange, forked; in humus or on rotting wood ..
....***Hygrophoropsis aurantiaca*** (141)

17b. Not with all the above characteristics............................... 18

18a. Cap white with brown scales; gill edges serrated; fruiting singly or in clusters on conifer logs or at the base of dead conifers ....***Neolentinus ponderosus*** (142)

18b. Cap not conspicuously scaly; gill edges not serrated 19

19a. Fruiting body orange but usually with olive tints; stalk central to off-center; growing in clusters on hardwood trees, stumps, and roots ....***Omphalotus olivascens*** (143)

19b. Not with all the above characteristics............................... 20

20a. Gills typically decurrent; cap often funnel shaped; spores white or tinged pink... 21

20b. Gills typically notched or adnexed; spores white............. 22

21a. Cap and spores whitish, or if cap brown or purple, then spores tinged pink; odor variable ....
.... see key to ***Clitocybe* spp.** (144)

21b. Cap white or tinged gray and spores pink; odor strongly mealy...***Clitopilus prunulus*** (156)

22a. Cap brownish, convex, often with a broad umbo; stalk stiff and straight; on ground in woods and landscaped areas ....
....***Melanoleuca melaleuca*** (157)

22b. Not with all the above characteristics; in woods.............. 23

23a. Cap and stalk yellow but covered with red fibrils; flesh and gills yellow; growing on or near wood...
....***Tricholomopsis rutilans*** (158)

23b. Not with all the above characteristics.....................................
....see key to ***Tricholoma* spp.** (159)

24a. Cap generally less than 2.5 cm broad, reddish brown; gills adnate to decurrent and well spaced; stalk tough, brownish........
....***Xeromphalina* spp.** (170)

24b. Not with all the above characteristics................................ 25

25a. Cap funnel shaped, yellow, pink, or orange; gills decurrent
.... ***Lichenomphalia umbellifera*** (171)

25b. Not with all the above characteristics; gills usually adnexed to adnate (but sometimes slightly decurrent)............................ 26

26a. Growing on conifer cones; gills not edged pink....................
.... ***Strobilurus trullisatus*** (172)

26b. Not growing on conifer cones, or if on cones, then gills edged pink 27

27a. Cap conical and umbonate, yellow-brown; stalk cartilaginous, long, tapered, and deeply rooted; fruiting under trees, especially common under coastal redwoods.....................................
....***Caulorhiza* spp.** (173)

27b. Not with all the above characteristics................................ 28

28a. Cap conical or bell shaped, at least when young, often translucent-striate; stalk thin and usually long, hollow, and fragile; dry fruiting bodies not reviving when wet ....
....see key to ***Mycena* spp.** (174)

28b. Cap usually convex to plane, striate or not; stalk often tough; dry fruiting bodies often reviving when wet 29

29a. Cap white; gills widely spaced; stalk white with a blackish base; growing on sticks ....***Marasmiellus candidus*** (183)

29b. Not with all the above characteristics................................ 30

30a. Cap brownish; the lower part of the stalk hairy; fruiting in dense clusters or in large groups .... ***Gymnopus* spp.** (185)

30b. Not with all the above characteristics................................ 31

31a. Cap chestnut brown and smooth; stalk cartilaginous; fruiting body not reviving in water after desiccation ....
.... ***Gymnopus dryophilus*** (185)

31b. Cap variously colored; stalk tough and polished or thin and wiry; fruiting body reviving in water after desiccation..........
....see key to ***Marasmius* spp.** (185)

Gilled Mushrooms with Reddish Spores

1a. Universal veil present, leaving a volva ....
....***Volvopluteus gloiocephalus*** (192)
1b. Universal veil absent 2
2a. Stalk absent; cap fan shaped and orange; on wood ....
.... ***Phyllotopsis nidulans*** (192)
2b. Stalk present 3
3a. Gills free; partial veil absent; fruiting on wood or ground ....
.... ***Pluteus* spp.** (194)
3b. Gills attached to the stalk; fruiting on ground 4
4a. Stalk fleshy; cap moderately sized to large (up to 12 cm or
more broad) .... ***Entoloma* spp.** (195–198)
4b. Stalk slender and cartilaginous; cap typically less than 7 cm
broad 5
5a. Stalk hollow and easily splitting; cap smooth, grayish or
brownish, usually umbonate ....
..............................***Entoloma* spp. subgroup *Nolanea*** (198–200)
5b. Not with all the above characteristics 6
6a. Cap and stalk blue to black ....
.... ***Entoloma* spp. subgroup *Leptonia*** (200)
6b. Cap and stalk white ....***Entoloma sericellum*** (201)

Gilled Mushrooms with Black Spores

1a. Gills decurrent 2
1b. Gills free to adnate 3
2a. Cap viscid to dry; flesh yellowish to pale orange ....
.... ***Chroogomphus* spp.** (201)
2b. Cap very viscid; flesh whitish to grayish ....
....***Gomphidius* spp.** (202)
3a. Gills usually autodigesting; fruiting on various substrates ....
.... see key to **Inky caps** (204)
3b. Gills not autodigesting; often on dung ....
.... ***Panaeolus* spp.** (212)

Gilled Mushrooms with Brown, Purple-Brown, or Purple-Black Spores

1a. Spores purple-brown or purple-black 2
1b. Spores dull brown, deep brown, or chocolate brown 7
2a. Cap brown, gray, or dull olive; flesh often bruising blue ....
....***Psilocybe* spp.** (213)

14b. Not with all the above characteristics................................ 15

15a. Cap usually viscid; gill edges minutely scalloped; apex of stalk powdery; odor radishlike ....***Hebeloma* spp.** (239)

15b. Cap dry, often umbonate or with a margin that splits, or both; surface usually silky, hairy, or scaly; odor sometimes like green corn..see key to ***Inocybe* spp.** (240)

Gilled Mushrooms with Orange-Brown to Rusty Brown Spores

1. Growing on wood (including wood chips)..............................2
1. Growing on the ground...4

2a. Cap up to 30 cm broad, yellow-orange, clustered on wood .. .. ***Gymnopilus junonius*** (245)

2b. Cap typically less than 10 cm broad 3

3a. Cap orange, up to 8 cm broad; ring on stalk a few hairs or absent ....***Gymnopilus sapineus*** (246)

3b. Cap brownish, less than 4 cm broad; ring on stalk usually present...***Galerina marginata*** (247)

4a. Fruiting body small, fragile, often withering quickly; cap conical or bell shaped; in grass, gardens, or woods 5

4b. Fruiting body not withering quickly 6

5a. Cap yellow; fruiting in grass or on dung.................................. .... ***Bolbitius titubans*** (248)

5b. Cap cream colored to brown; fruiting in grass .... ***Conocybe* spp.** (249)

6a. Partial veil absent; cap often conical or umbonate, usually viscid; stalk with a long, tapered, rooting base............................... ....***Phaeocollybia* spp.** (250)

6b. Not with all the above characteristics; partial veil absent or present..7

7a. Veil evanescent, sometimes leaving whitish flecks on the cap..***Tubaria* spp.** (251)

7b. Veil cobweblike, often leaving hairs on the stalk ....see key to ***Cortinarius* spp.** (252)

Nongilled Mushrooms

1. Fruiting body with a cap (often vase shaped) and stalk; underside of cap wrinkled or with shallow folds and ridges............. ....**Chanterelles and similar fungi** (270–277)

1. Spores borne in pores (sometimes gill-like), on spines, on a smooth surface, etc.; cap present or absent...................................2

2a. Fruiting body with a layer of downward-pointing spines or teeth on the underside of a cap or cushion ....
.... see key to **Tooth fungi** (277)

2b. Fruiting body without spore-bearing spines or teeth...........3

3a. Fruiting body a mass of ribbonlike, cream-colored branches; fruiting at the base of conifers.........***Sparassis radicata*** (287)

3b. Not with all the above characteristics....................................4

4a. Fruiting body clublike or coral-like ....
.... see key to **Club and coral fungi** (288)

4b. Fruiting body not clublike or coral-like 5

5a. Fruiting body gelatinous, jellylike, or rubbery, usually growing on wood... see key to **Jelly fungi** (303)

5b. Fruiting body not gelatinous, jellylike, or rubbery...............6

6a. Pores absent 7

6b. Pores present (sometimes gill-like) 9

7a. Cap fanlike, upright in clusters, brownish ....
....***Thelephora* spp.** (310)

7b. Cap bracketlike or absent...8

8a. Fruiting body thin, leathery, bracketlike (but sometimes partially flat on substrate)................................***Stereum* spp.** (311)

8b. Fruiting body flat on substrate ....***Phlebia radiata*** (312)

9a. Tubes separable from cap; stalk typically central.............. 10

9b. Pores inseparable from cap; stalk present and central or lateral or absent.. 14

10a. Stalk roughened by small, dark tufts of hairs ....
.... ***Leccinum manzanitae*** (313)

10b. Stalk without tufts of hairs... 11

11a. Cap, stalk, and pores brown; spore print reddish brown.....
.... ***Porphyrellus porphyrosporus*** (315)

11b. Fruiting body not completely brown; spores usually olive brown or brown 12

12a. Cap up to 7 cm broad, orange-brown, taste peppery...........
.... ***Chalciporus piperatus*** (316)

12b. Not with all the above characteristics................................ 13

13a. Cap usually dry; pores white, yellow, red, brown, or gray, sometimes bluing when injured; stalk fleshy, sometimes bulbous, reticulate, or both; partial veil absent....................................
.... see key to **Boletus spp.** (317)

13b. Cap usually viscid; pores white, yellow, rarely red, often radially arranged; stalk equal, resinous glandular dots or smears often present, partial veil often present ....
....see key to **Suillus spp.** (328)

14a. Spore-bearing surface composed of separate tubes; fruiting body fleshy, often exuding red water droplets when moist...........
....***Fistulina hepatica*** (335)
14b. Spore-bearing surface composed of pores........................ 15
15a. Stalk present; annual; on wood or the ground................. 16
15b. Stalk absent; annual or perennial; on wood..................... 19
16a. On wood; cap brownish to black; stalk partly black.............
....***Polyporus varius*** (336)
16b. On ground (rarely on wood)... 17
17a. Fruiting from an underground tuber; cap tan to brown......
....***Polyporus tuberaster*** (337)
17b. Underground tuber absent.. 18
18a. Cap dark brown, velvety, taste bitter.................................
....***Jahnoporus hirtus*** (338)
18b. Cap differently colored, or if brown, then not bitter tasting...***Albatrellus ovinus*** (339)
19a. Pore surface completely enclosed by a tough membrane ....
....***Cryptoporus volvatus*** (341)
19b. Pore surface exposed... 20
20a. Fruiting body usually flat on substrate............................. 21
20b. Fruiting body normally with a cap..................................... 22
21a. Fruiting body flat and thin, tightly attached to logs............
...................pores white to cream colored: ***Antrodia xantha*** (342);
....pores orange: ***Ceriporia spissa*** (343);
.....................................or pores pink: ***Oligoporus placentus*** (347)
21b. Fruiting body up to 3 cm thick, orange, soft and spongy, pore walls toothlike in age......... ***Pycnoporellus alboluteus*** (343)
22a. Spore-bearing surface consisting of mazelike pores or platelike gills................................... ***Gloeophyllum sepiarium*** (344)
22b. Spore-bearing surface consisting of pores........................ 23
23a. Fruiting body corky, hard, or woody, medium sized to very large and thick; annual or perennial........ see key to **Conks** (345)
23b. Fruiting body thin and leathery or thick and soft and spongy.. 24
24a. Cap soft, watery, spongy, pore surface usually whitish........
.... ***Oligoporus leucospongia*** (354)
24b. Cap thin, tough, and leathery, typically tearing easily; shelflike; pores tiny... 25
25a. Cap white to gray with a purple margin, pores purplish.....
....***Trichaptum abietinum*** (355)
25b. Cap conspicuously zoned with alternating bands of brown, red, and gray... ***Trametes versicolor*** (356)

Gasteroid fungi

1. Fruiting body a small funnel-shaped cup less than 1 cm broad containing one or more small egglike spore cases ....
.... **Nidula candida** (357)
1. Not with all the above characteristics.....................................2
2a. Fruiting body with a volva, stalk, and head, the latter covered with a slimy, foul-smelling spore mass.................................
....**Phallus impudicus** (358)
2b. Not with all the above characteristics.....................................3
3a. Fruiting body with a stalk below the round spore case........4
3b. Stalk absent or rudimentary or present simply as a sterile base...5
4a. Fruiting body with a puffball-like spore case that sits on a well-developed stalk..........................**Tulostoma campestre** (359)
4b. Stalk of fruiting body reaching top of spore case that contains contorted plates ....**Agaricus deserticola** (360)
5a. Fruiting above ground; spore case rupturing at maturity, spore mass powdery when mature; fruiting body usually round to oval or outer skin splitting into starlike rays ....
....see key to **Puffballs and earthstars** (361)
5b. Fruiting below ground; spore case remaining intact a long time; spore mass not powdery at maturity......................................
.... **Rhizopogon spp.** (377) and **Hydnangium carneum** (378)

Ascomycetes

1. Asci borne in minute flasks (perithecia) produced in a firm fruiting body 2
1. Asci typically borne in a layer on or in a fruiting body 5
2a. Growing on wood 3
2b. Not growing on wood..4
3a. Fruiting body black and rounded ....
.... **Annulohypoxylon thouarsianum** (379)
3b. Fruiting body antlerlike.............................. **Xylaria spp.** (380)
4a. Growing on insects or truffles; fruiting body clublike ....
........................**Cordyceps** and **Elaphocordyceps spp.** (381–382)
4b. Growing on mushrooms and assuming their shape..............
....**Hypomyces spp.** (382)
5a. Fruiting body completely or partly subterranean 6
5b. Fruiting body on soil, duff, or wood.....................................7
6a. Fruiting body opening at soil surface at maturity..................
.... inner surface purple: **Sarcosphaera coronaria** (383);

.... or yellowish: **Geopora arenicola** (383)

6b. Fruiting body subterranean and tuberous, never releasing spores at the soil surface..
.... interior marbled and firm: **Tuber spp.** (384);
.... or interior hollow: **Genea arenaria** and others (385)

7a. Fruiting body generally stalked..8

7b. Fruiting body generally sessile 15

8a. Fruiting body cuplike, the stalk central and not running up the side of the fruiting body..
.... cup scarlet red: **Sarcoscypha coccinea** (386);
.... black: **Plectania spp.** (387);
.... bright orange: **Sowerbyella rhenana** (387);
....or brownish: **Geopyxis carbonaria** (388)

8b. Fruiting body variously shaped; if cuplike, ribs of the stalk running up the side of the cup..9

9a. Fruiting body clublike or spoonlike, lacking a distinct cap...
....**Trichoglossum hirsutum** (389)

9b. Fruiting body with a distinct cap or cuplike 10

10a. Flesh gelatinous; cap rounded, viscid, often brightly colored...**Leotia spp.** (391)

10b. Flesh not gelatinous 11

11a. Cap honeycombed with ridges and pits; most or all of cap fused with stalk..**Morchella spp.** (392)

11b. Cap smooth, wrinkled, brain, cup, or saddle shaped...... 12

12a. Cap smooth to wrinkled, attached only to very top of stalk, the sides hanging free...**Verpa spp.** (393)

12b. Cap brain, cup, or saddle shaped 13

13a. Cap brainlike or wrinkled.................. **Gyromitra spp.** (394)

13b. Cap cup or saddle shaped 14

14a. Cap cuplike, the stalk with ribs that run up the side of the cup ....**Helvella acetabulum** (395)

14b. Cap convoluted or saddle shaped ....
.... **Helvella lacunosa** and others (396)

15a. Fruiting on wood 16

15b. Fruiting on the ground.. 22

16a. Fruiting tough, gelatinous, gumdroplike, blackish..............
....**Bulgaria inquinans** (397)

16b. Fruiting body fleshy and fragile, cuplike or disklike, often brightly colored.. 17

17a. Fruiting body bright red.. 18

17b. Fruiting body not bright red... 19

18a. Fruiting body saucerlike or disklike, typically less than 1 cm broad, the margin decorated with dark hairs..........................
....***Scutellinia* spp.** (398)
18b. Fruiting body a cup typically larger than 1 cm, margin hairless .... ***Sarcoscypha coccinea*** (386)
19a. Fruiting body generally larger than 2 cm broad, cup shaped becoming nearly flat, often brown ....***Peziza* spp.** (399)
19b. Fruiting body generally smaller than 2 cm broad 20
20a. Fruiting body reddish purple, cup or disk shaped...............
.... ***Ascocoryne sarcoides*** (400)
20b. Fruiting body yellow... 21
21a. Fruiting body cuplike, yellow, fringed with short hairs ....
.... ***Lachnellula arida*** (401)
21b. Fruiting body disklike, yellow, hairs absent...........................
.... ***Bisporella citrina*** (402)
22a. Fruiting body an erect asymmetrical cup slit down one side..***Otidea* spp.** (403)
22b. Fruiting body cup shaped or flattened and spreading at maturity 23
23a. Fruiting body brown, flattened, growing in the mountains following snowmelt ....***Discina perlata*** (404)
23b. Fruiting body bright orange, cup shaped or flattened at maturity 24
24a. Fruiting body staining bluish green; growing in the mountains following snowmelt......................***Caloscypha fulgens*** (405)
24b. Fruiting body not staining bluish green; widespread...........
.... ***Aleuria aurantia*** (406)

BASIDIOMYCETES

Amanita Species

These are some of the best known and most dangerous mushrooms. They are medium to large, often colorful, and common in forests in the western United States. White spores, gills free or narrowly attached to the stalk, the presence of a universal veil, and mycorrhizal association with oaks, birches, and conifers are the principal field marks. The universal veil often leaves warts or patches of tissue on the cap surface and a volva at the base of the stalk. The volva may be collarlike, scaly, or saclike (attached just to the base of the stalk). A ring on the stalk is present or absent. Although some *Amanita* species are edible, beginning mushroom hunters should follow the rule that all amanitas are off-limits for the table because an error in identification could be fatal!

a. Volva collarlike or scaly
- b. Cap and stalk developing a pinkish blush ....***Amanita novinupta***
- b. Cap and stalk not developing a pinkish blush, but cap may be red
  - c. Partial and universal veils yellow; cap brownish.............. ....***A. franchetii***
  - c. Partial veil not yellow, or if yellow, then cap red or orange
    - d. Cap white and covered with patches of white universal veil tissue...***A. silvicola***
    - d. Cap cream colored or yellow; warts white.................... .... ***A. gemmata***
    - d. Cap brown; warts white .... ***A. pantherina***
    - d. Cap red; warts white or yellowish..............***A. muscaria***

a. Volva saclike or constricted around the stalk
- e. Ring absent; cap conspicuously striate

f. Volva fused around the stalk with the top margin flaring away..................................**A. constricta** (see *A. vaginata*)

f. Volva attached only at the bottom of the stalk

  g. Cap orangish or pink..**A. velosa**

  g. Cap brown...**A. pachycolea**

  g. Cap gray ....**A. vaginata**

e. Ring present; cap margin smooth or striate

  h. Cap white, margin usually not striate..............**A. ocreata**

  h. Cap yellow-green or yellow

    i. Cap yellow-green, the margin usually not striate, cap often bald ....**A. phalloides**

    i. Cap yellow, the margin striate, cap often partially covered with a thick cottony patch...........**A. vernicoccora**

  h. Cap orange to orange-brown

    j. Stalk smooth, cap partially covered with thick cottony patch ....**A. calyptroderma**

    j. Stalk fibrillose, cap usually bald.................**A. jacksonii**

Amanita calyptroderma

(Other names: *Amanita calyptrata, A. lanei*) Coccora

CAP: 10 to 25 cm in diameter, convex becoming plane, orange to orange-brown with a paler striate margin, surface smooth with a thick white cottony patch resembling meringue on top, flesh white, odor mild. **GILLS:** Free, close, and white to creamy. **STALK:** 10 to 20 cm long, 2 to 4 cm wide, equal, white or creamy, smooth, hollow, usually with a skirtlike pale yellow ring, which is fragile and may disappear. **VOLVA:** White, thick, cottony, and saclike. **SPORES:** White, elliptical, smooth, inamyloid. **HABITAT:** Single to gregarious on ground in mycorrhizal association with trees in mixed woods, especially common in fall. **EDIBILITY:** For some, edible and choice, but the risks of confusion with deadly *Amanita* species are so serious that only the most experienced collector should consider it for the table.

This common and robust mushroom of coastal and montane forests is prized by many collectors, but caution is absolutely essential since young, old, or atypical specimens may be confused with poisonous species. The cap is typically orange-brown, but white forms are occasionally encountered. ***Amanita vernicocco-***

Amanita calyptroderma

ra (p. 82), a closely related species, fruits in the hills and mountains in late winter and spring. It is similar to *A. calyptroderma*,
but the cap is dull pale yellow. It also has a cottony universal veil
that leaves a patch on the cap and a thick volva.

Amanita franchetii
(Other name: *Amanita aspera*)
CAP: 5 to 12 cm in diameter, convex becoming plane, brown to
yellow-brown with yellow to pale yellow-brown warts scattered
over the cap surface, the margin smooth or faintly striate in age,
flesh white to pale yellow, odor mild. **GILLS:** Free, close, white or
pale yellow. **STALK:** 6 to 15 cm long, 1 to 2 cm wide, equal or tapering upward, often with an enlarged or bulbous base, white or
tinged yellow, with a fragile, skirtlike, pale yellow ring with a
brighter yellow margin. **VOLVA:** Pale yellow concentric rings on
the basal bulb. **SPORES:** White, elliptical, smooth, amyloid. **HABI-
TAT:** Single or in small groups in mycorrhizal association with
trees in mixed hardwood/coniferous forests. **EDIBILITY:** Unknown;
possibly toxic.

This attractive *Amanita* is recognized by the yellow color of
the partial and universal veils. The persistently yellow margin of
the ring helps identify this mushroom even when the brown

Amanita franchetii

background color of the cap fades in age. It often fruits in the same location year after year in mycorrhizal association with various hardwoods and conifers. Because *A. franchetii* is a close relative of **A. muscaria** (p. 72) and **A. pantherina** (p. 76), both which contain toxins, it is probably unsafe to eat.

Amanita gemmata
Gemmed Amanita

CAP: 4 to 16 cm in diameter, rounded or convex, becoming almost plane, the margin incurved at first, smooth or striate, the surface cream colored, dull yellow, or light yellow-orange with white to cream warts that may wear off, flesh white, odor mild. **GILLS:** Adnexed to free, white. **STALK:** 5 to 15 cm long, 1 to 2 cm wide, equal or tapered upward, white or creamy, smooth above and fibrillose below the white membranous ring, which may

Amanita gemmata

disappear. **VOLVA:** White, cottony, collarlike, tightly adhering to the stalk with a free rim. **SPORES:** White, elliptical, smooth, inamyloid. **HABITAT:** Single or in small groups in coastal and montane mixed forests. **EDIBILITY:** Poisonous; this species is closely related to *A. muscaria* and *A. pantherina*, which are also toxic.

The size and color of the cap of *A. gemmata* are variable. The cap typically is cream colored, but it sometimes appears bright yellow. In addition, the whitish warts or patches on the cap may fall off, leading to confusion with other species. Some researchers speculate that it may represent a complex of species. **Amanita aprica** is a bright yellow or orange species that commonly fruits in spring or summer in the Sierra Nevada. It has warts on the cap that cannot be washed off, unlike the warts of *A. gemmata*.

Amanita jacksonii
(Other name: *Amanita caesarea*) Caesar's Mushroom
CAP: 6 to 20 cm in diameter, initially oval becoming convex, bright orange or sometimes red, fading to yellow in age, smooth

or sometimes with a patch of white universal veil tissue, viscid when moist, the margin striate about halfway to the center, flesh whitish to pale yellow, not changing when bruised. **GILLS:** Adnexed to free, bright yellow. **STALK:** 6 to 20 cm long, 1 to 3 cm wide, equal, yellow or pale yellow, often with orange fibrils arranged in zones, partial veil leaving an orange or yellow persistent, membranous skirtlike ring. **VOLVA:** White, thick, and saclike, attached only at the base. **SPORES:** White, elliptical, smooth, inamyloid. **HABITAT:** Single or scattered in mycorrhizal association with oaks and pines, common in the mountains of Arizona and New Mexico after summer monsoon rains. **EDIBILITY:** Edible, but caution is advised to avoid confusion with *A. muscaria* and *A. phalloides*, the Death Cap.

Like many North American mushrooms that were originally named after their European counterparts, the name of the western counterpart of Caesar's Mushroom is often debated, especially since it probably represents a complex of species. The bright orange or red striate cap, yellow gills and stalk, and white universal veil tissue make a striking color combination.

Amanita jacksonii

Amanita muscaria

Amanita muscaria **(pp. 412–413)**

Fly Agaric

CAP: 8 to 30 cm in diameter, rounded and then convex and finally plane, typically red to orange-red but sometimes yellow or white, with white or pale yellow warts, the margin usually striate. **GILLS:** Adnexed to free, close, white. **STALK:** 6 to 16 cm long, 1 to 3 cm wide, tapering upward, white, smooth above the skirtlike ring, fibrillose below. **VOLVA:** White to pale yellow, collarlike, consisting of one or more concentric rings around the lower stalk and top of the basal bulb. **SPORES:** White, elliptical, smooth, inamyloid. **HABITAT:** Single, scattered, or in large groups and rings in mycorrhizal association with many trees, especially pines. **EDIBILITY:** Poisonous and hallucinogenic; the toxins are water soluble, but given the preparation required to remove the toxins, this is not a good mushroom for the table.

Amanita muscaria is one of the most readily recognized mushrooms, found in all manner of artwork and home decorations. The defining features include the warts on the cap, the ring on the stalk, and the concentric rinds of universal veil tissue on the enlarged base of the stalk. The red form is the most common, but specimens with yellow or even paler caps are sometimes encountered. Its popularity includes tales fanciful to factual. The name Fly Agaric refers to the fact that flies become quickly inebriated after drinking a concoction of a piece of the mushroom in milk or some sugary liquid.

Amanita novinupta

(Other name: *Amanita rubescens*) Blusher

CAP: 5 to 15 cm in diameter, convex becoming plane, white with unevenly distributed pink or pinkish brown areas, the surface decorated with white, pink, or reddish brown scattered warts, the margin usually not striate. **GILLS:** Adnexed to free, white or off-white in age. **STALK:** 5 to 15 cm long, 1 to 3 cm wide, tapered upward with an enlarged base, white with pink or pale pinkish brown fibrils below the similarly colored skirtlike ring. **VOLVA:** Collarlike with one to several concentric rings. **SPORES:** White, elliptical, smooth, amyloid. **HABITAT:** Single or scattered under trees, especially oaks. **EDIBILITY:** Unknown; extreme caution should be exercised when collecting any species of *Amanita*.

This species of *Amanita* is easily identified by the pink "blushing" of the cap and stalk. It was long misidentified as *A. rubescens*, an eastern U.S. species. *Amanita novinupta* is related to **A. franchetii** (p. 68), which has yellow-tinged universal veil tissue rather than pink. Specimens that have aged brown may

Amanita novinupta

resemble *A. pantherina* (p. 76), and specimens growing in the direct sun may be whiter than normal, possibly causing confusion with *A. magniverrucata*, an all-white species with distinctive large, erect, pyramidal warts. None of these mushrooms should be considered edible.

Amanita ocreata
Destroying Angel

CAP: 5 to 15 cm in diameter, convex becoming plane, sometimes shallowly depressed in age, white but often developing yellow to brown stains, usually bald but sometimes with a thin patch of universal veil tissue, the margin typically not striate. **GILLS:** Adnexed to free, close, white. **STALK:** 6 to 20 cm long, 1 to 3 cm wide, white, more-or-less equal with an enlarged base and a thin, fragile, white ring that sometimes disappears in age. **VOLVA:** Saclike, white, membranous. **SPORES:** White, elliptical, smooth, amyloid. **HABITAT:** Single or scattered, mycorrhizal with trees, especially oaks, typically fruiting in spring. **EDIBILITY:** Deadly poisonous; confusion with young Meadow Mushrooms (*Agaricus campestris*) is possible if the saclike volva goes unnoticed.

The Destroying Angel and its relative the Death Cap, *A. phalloides* (p. 77), are deadly poisonous mushrooms frequently encountered under oaks in woods, parks, and gardens. Because poisonings due to these species occur nearly every year in the western United States, the identifying characteristics of these species must be known by all collectors of mushrooms. The white spores, free or nearly free gills, partial and universal veils, and mycorrhizal habit are key characteristics. *Amanita ocreata* represents a complex of two or three species in the western United States. All typically fruit in spring.

Amanita pachycolea
Western Grisette

CAP: 8 to 20 cm in diameter, convex or oval, becoming plane in age, dark brown aging pale grayish brown, viscid when moist, bald or rarely with patches of white universal veil tissue, the margin conspicuously striate. **GILLS:** Adnexed to free, white but sometimes with brown edges. **STALK:** 10 to 25 cm long, 1 to 3 cm wide, equal or tapering upward, white with tiny brown scales,

Amanita ocreata

Amanita pachycolea

partial veil absent. **VOLVA:** White, saclike, large and loose, only attached at the base of the stalk but extending up the stalk, often stained reddish brown. **SPORES:** White, globose to subglobose, smooth, inamyloid. **HABITAT:** Single or scattered in coastal mixed hardwood/coniferous forests. **EDIBILITY:** Edible but extreme caution must be exercised since it might be confused with atypical specimens of poisonous species of *Amanita.*

The dark brown color (at least when young) and deep striations of the cap, ringless stalk, large sheathing volva with rusty spots, and relatively tall stature of this mushroom separate it from its grisette cousins, *A. constricta* (p. 80) and *A. vaginata* (p. 79). Like other species of *Amanita*, it is mycorrhizal with various conifers and hardwoods. The rich color of the smooth cap and the crisp and clean look, which is typical of *Amanita* species, make this one of the most attractive mushrooms of our coastal forests.

Amanita pantherina
Panther Cap

CAP: 5 to 18 cm in diameter, convex becoming plane in age, dark brown but sometimes hazel to yellowish brown, the margin with short striations, warts loose, white to buff. **GILLS:** Adnexed to free, close, white but off-white to yellowish in age. **STALK:** 5 to 15 cm or more long, up to 3 cm wide, equal or enlarged downward to a basal bulb, white, finely fibrillose below the white, skirtlike ring. **VOLVA:** White, collarlike with a free rim adhering to the bulbous base. **SPORES:** White, elliptical, smooth, inamyloid. **HABITAT:** Single or scattered on ground in mycorrhizal association with pines; widespread but especially common in spring in the high mountains under various conifers. **EDIBILITY:** Poisonous.

The brownish cap, white warts, and collarlike volva are the helpful field marks. The above description fits a complex of at least two species in California (neither of which may be the true *A. pantherina* of Europe). *Amanita pantherina* is related to *A. gemmata* (p. 69) and *A. muscaria* (p. 72). Pale-colored specimens of *A. pantherina* might be confused with *A. gemmata*, but the cap of *A. gemmata* is generally more yellow or cream colored. All these species all contain toxins and should be avoided.

Amanita pantherina

Amanita phalloides

Death Cap

CAP: 5 to 15 cm in diameter, oval and then convex to nearly plane, greenish yellow, yellow, or sometimes light brownish green, fading to yellow or almost white, the margin typically not striate, the surface usually bald but sometimes with one or more patches of thin, white universal veil tissue. **GILLS:** Free, close, white. **STALK:** 5 to 18 cm long, 1 to 3 cm wide, more-or-less equal with an enlarged base, white, the partial veil membranous, leaving a large skirtlike ring high on the stalk. **VOLVA:** Saclike, white, large, and membranous with a free margin. **SPORES:** White, elliptical, smooth, amyloid. **HABITAT:** Single or scattered on the ground in mycorrhizal association with a number of trees, especially live oak. **EDIBILITY:** Deadly poisonous.

The Death Cap is probably responsible for most mushroom fatalities in the western states. Because the color of its cap can be quite variable, mushroom hunters need to be able to identify it based on a combination of features: white spores, universal veil leaving a volva (be sure to check the stalk below ground since

Amanita phalloides

the volva can be concealed), greenish cast of the cap, membranous ring, and association with trees, especially oaks. The cap is often a shiny yellow-green, but it sometimes has brownish tints and can be whitish if growing in the direct sunlight. Evidence suggests it is an introduced species that is continuing to spread in the West.

Amanita silvicola

CAP: 5 to 12 cm in diameter, convex becoming plane in age, the margin incurved and often edged with veil tissue, the surface uniformly white, covered completely or with patches of thick, soft universal veil tissue that may wash off in rain, odor mild. **GILLS:** Adnexed to free, white. **STALK:** 6 to 10 cm long, 1 to 2.5 cm wide, equal or tapering upward, white, the base somewhat bulbous and rounded, partial veil leaving a ring that may disappear. **VOLVA:** White, collarlike or consisting of rings of white tissue on the base of the stalk or extending up the stalk, sometimes stained reddish brown. **SPORES:** White, elliptical, smooth, amyloid. **HABITAT:** Single or scattered in coastal mixed hardwood/coniferous forests. **EDIBILITY:** Avoid; possibly toxic.

Amanita silvicola

Amanita silvicola is uniformly white with soft and cottony universal veil tissue on the top of cap. The bulb at the base of the stalk is not deeply seated in the soil, like the similar **A. smithiana**, which has a white cap with soft cottony warts. In contrast to *A. silvicola*, *A. smithiana* may have an unpleasant odor. **Amanita magniverrucata** is an all-white species with distinctive large, erect, pyramidal warts. Like *A. smithiana*, it has an extended rootlike stalk base. **Amanita baccata**, another white species with a long-rooting stalk base, has flattened veil tissue on the cap. It prefers sandy habitats.

Amanita vaginata

Grisette

CAP: 5 to 10 cm in diameter, oval to convex becoming plane, gray to grayish brown, bald or occasionally with a patch of white universal veil tissue that may discolor yellowish, the margin striate, odor mild. **GILLS:** Adnexed to free, white, close. **STALK:** 6 to 15 cm long, 1 to 2 cm wide, white or tinged gray with light gray appressed fibrillose scales, partial veil absent. **VOLVA:** Saclike, membranous, attached only at the base of the stalk, sometimes stained with yellowish or rusty spots. **SPORES:** White, globose to subglobose, smooth, inamyloid. **HABITAT:** Single or scattered on

Amanita vaginata

ground under hardwoods and conifers. **EDIBILITY:** Edible, but extreme caution must be exercised since it might be confused with atypical specimens of extremely poisonous species of *Amanita*.

Amanita vaginata is one of the grisettes, which includes **A. constricta** and **A. pachycolea** (p. 74). All have strongly striate caps and ringless stalks. *Amanita vaginata* is the smallest of the group, with more gray and less brown tones in the cap than the other two. Unlike *A. vaginata*, *A. constricta* has a volva that adheres tightly to the stalk before flaring out. **Amanita protecta** has a grayish brown cap, and a volva that breaks apart, leaving bands of gray fibrils on the stalk. Although these species do not contain any known toxins, they are not good candidates for the table because some *Amanita* species are deadly poisonous.

Amanita velosa
Spring Amanita
CAP: 5 to 15 cm in diameter, convex and then plane, orange-pinkish to salmon-buff, the margin conspicuously striate, the surface usually with a substantial patch of white universal veil tissue. **GILLS:** Adnexed to free, close, white but sometimes with a

dull pink tinge in age. **STALK:** 5 to 15 cm long, 1 to 3 cm wide, equal, whitish, partial veil absent, but a faint discolored zone may be present where the ring would be expected. **VOLVA:** Saclike, white, sheathing the base of the stalk. **SPORES:** White, elliptical, smooth, inamyloid. **HABITAT:** Single to scattered in spring, often fruiting in surprisingly dry habitats but always in mycorrhizal association with trees, especially oak. **EDIBILITY:** Edible, but this species grows under oaks at the same time as the deadly *A. ocreata*, so great care must be exercised if collected for the table.

This mushroom is a bellwether of spring. *Amanita velosa* is related to the grisettes, **A. constricta** (p. 80), **A. pachycolea** (p. 74), and **A. vaginata** (p. 79) and, like them, has a striate cap, ringless stalk, and saclike volva. The orange-buff or pinkish color of the cap and the presence of a white patch of universal veil tissue on the surface of the cap help identify this species. Of the group, *A. velosa* is the best species to eat, but only experienced collectors should collect any *Amanita* species for the table.

Amanita velosa

Amanita vernicoccora
Spring Coccora

CAP: 6 to 18 cm in diameter, convex becoming plane, uniform yellow to pale yellow, surface smooth with a central white cottony patch, margin short striate, flesh white to pale yellow, odor and taste mild to pungent. **GILLS:** Free, crowded, white to creamy. **STALK:** 5 to 14 cm long, 1.5 to 3 cm wide, equal, white or creamy, smooth, hollow to stuffed with a cottony or jellylike substance, usually with a skirtlike white to pale yellow ring that collapses in age. **VOLVA:** White, thick, saclike. **SPORES:** White, elliptical, smooth, inamyloid. **HABITAT:** Single to scattered on ground in mycorrhizal association with a variety of trees, especially black oak (*Quercus kelloggii*) in the Sierra Nevada; fruiting in late winter and spring. **EDIBILITY:** Edible, but confusion with deadly *Amanita* species is possible.

This endemic California species is edible, but because confusion with toxic species of *Amanita* is possible, it is not recommended for the table. *Amanita vernicoccora* is characterized by its yellow cap, thick remnant of cottony universal veil tissue on the cap, and springtime fruiting habit. The Coccora, **A. calyptroderma** (p. 67), is a close relative and for a long time was considered conspecific with *A. vernicoccora*. *Amanita calyptroderma* has a bright orange to orange-brown cap and cottony universal veil that leaves a thick volva and cap remnant. It fruits in fall and winter.

Amanita vernicoccora

Lepiota and Allies

This assemblage of genera and species has white or green spores, free gills, and a partial veil that leaves a membranous or fragile ring on the stalk. The mushrooms range from small and fragile to large and robust. The microscopic features of cross walls at the base of basidia or the cells on the surface of the caps must be examined to distinguish some species. All of the species are saprobic on decaying plant matter. This group has few safe edibles. Some people collect *Chlorophyllum brunneum* for the table, but adverse reactions have been reported. Otherwise, many of these species are very dangerous and should be strictly avoided. Some of the smaller species are known to contain the same deadly toxins present in *Amanita phalloides*, the Death Cap. *Chlorophyllum molybdites*, which is very common on lawns in warm locations, contains an unknown toxin that causes severe gastrointestinal distress. *Chlorophyllum molybdites* has been speculated to be the most commonly consumed poisonous mushroom in North America, possibly because it grows in home lawns and is morphologically similar to *C. brunneum*.

a. Spores green, growing in lawns and gardens..............................
.... ***Chlorophyllum molybdites***

a. Spores white, widespread
   b. Cap often more than 8 cm broad
      c. Cap smooth and white, stalk base enlarged......................
      ....***Leucoagaricus leucothites***
      c. Cap scaly, stalk base bulbous ....
      .... ***Chlorophyllum brunneum***
   b. Cap usually less than 8 cm broad
      d. Fruiting body yellow.............. ***Leucocoprinus birnbaumii***
      d. Fruiting body not yellow, often predominately white with dark center
         e. Stalk shaggy below the ring .... ***Lepiota magnispora***
         e. Stalk smooth
            f. Cap scales black or gray............... ***Lepiota atrodisca***
            f. Cap scales brown or reddish brown
               g. Cap scales concentrically arranged ....
        .... ***Lepiota castaneidisca*** (see *L. magnispora*)

g. Cap cuticle breaking up into radiating fibrils .....
....**Lepiota rubrotinctoides**

Chlorophyllum brunneum
Shaggy Parasol

CAP: 6 to 20 cm in diameter, rounded when young, becoming convex to nearly flat in age, white with large brown scales arranged more-or-less concentrically, the center usually brown, the surface dry, flesh bruising yellowish and finally reddish brown, odor and taste mild. **GILLS:** Free, white or aging brown. **STALK:** 5 to 20 cm long, 2 to 3 cm wide, equal with an abrupt basal bulb with a distinct rim, white, dry, bruising and aging orange to reddish brown, the partial veil leaving a double-edged, movable ring. **SPORES:** White, elliptical, smooth, dextrinoid. **HABITAT:** Single, in clusters, or in troops as a saprobe on ground in gardens and lawns, along roadsides, and under trees in landscaped areas. **EDIBILITY:** Edible and good, but some people are apparently allergic and have adverse reactions to it.

This is an easy mushroom to identify—the large brown scales on an otherwise white cap, free gills, white spores, presence of a ring, and stalk with a distinct basal bulb are distinguishing characteristics. **Chlorophyllum rachodes** (*Macrolepiota*

Chlorophyllum brunneum

rachodes) is similar, but it has a swollen base without a distinct rim. Both occur in the western United States and are edible, but caution is advised since some people apparently are allergic to them.

Chlorophyllum molybdites

CAP: 8 to 30 cm in diameter, bluntly conical when young, becoming convex to nearly flat, initially smooth and brown but the cuticle soon breaking up into brown scales or patches, revealing the white background color, the center remaining somewhat brown with progressively fewer brown scales toward the margin, the surface dry, odor and taste mild. **GILLS:** Free, white but becoming green as the spores mature. **STALK:** 5 to 20 cm long, 1 to 3 cm wide, equal or enlarged at base, white, dry, smooth, discoloring reddish brownish when bruised or in age, the partial veil leaving a membranous, thick, double-edged ring. **SPORES:** Green, elliptical, smooth, dextrinoid. **HABITAT:** Scattered or in fairy rings in grass; common in lawns in the San Joaquin Valley and southern California in summer and fall. **EDIBILITY:** Poisonous; the effects on children may be quite severe.

This mushroom has a unique feature—green spores. As the spores mature, the gills also become green. It is important to

Chlorophyllum molybdites

collect mature specimens with mature spores for accurate identification. It might be confused with **C. brunneum** (p. 84), but that species has white spores and the cap is more densely scaly. The distinction is important because *C. brunneum* is edible and *C. molybdites* will cause gastrointestinal distress. Immature specimens of the two species are virtually identical.

Lepiota atrodisca

CAP: 2 to 5 cm in diameter, convex becoming plane with a low umbo, whitish with gray to black scales, in age the center remaining blackish while the black scales fade toward the margin, the surface dry, not changing color when bruised. **GILLS:** Free, close, white. **STALK:** 3 to 9 cm long, 2 to 6 mm wide, equal, white, smooth or finely fibrillose toward the base, not changing color when bruised, the partial veil leaving a white fragile ring. **SPORES:** White, elliptical, smooth, dextrinoid. **HABITAT:** Single or scattered in mixed hardwood/coniferous forests, fruiting soon after the first fall rains along the California central and northern coast and in Oregon. **EDIBILITY:** Unknown; some small *Lepiota* species are dangerously poisonous.

The blackish center of the cap, black scales that fade toward the margin, free gills, white spores, and slender stalk define *L. atrodisca*, which represents a complex of several species. Many small woodland *Lepiota* species may be encountered, but the black cap scales and unchanging tissue when bruised makes this small mushroom distinctive. Like some other *Lepiota* species, it fruits after the first fall rains. None of these small species of *Lepiota* should be eaten because of the possible presence of amanitins, the same toxins present in **Amanita phalloides,** the Death Cap (p. 77).

Lepiota magnispora

(Other name: *Lepiota clypeolaria*)

CAP: 2 to 7 cm in diameter, blunt-conical becoming convex to nearly plane with a low umbo, the center smooth and dark brown, otherwise with yellow-brown to brown scales on a white background, pieces of the partial veil sometimes adhering to the margin, the surface dry. **GILLS:** Free, close, white when young, becoming creamy in age. **STALK:** 2 to 12 cm long, 3 to 8 mm

Lepiota atrodisca

wide, equal, white with yellowish shaggy scales below the fragile white ring, smooth or fibrillose above the ring. **SPORES:** White, spindle shaped, smooth, dextrinoid. **HABITAT:** Single or scattered on ground in mixed hardwood/coniferous forests. **EDIBILITY:** Unknown; some small *Lepiota* species are dangerously poisonous.

Lepiota magnispora

The white spores, free gills, absence of a universal veil, yellow-brown cap with a smooth brown center, and especially the shaggy stalk are distinguishing features of this woodland *Lepiota*. The overall light color of the cap with a center target is eye-catching. The cap of **L. castaneidisca** (*Lepiota cristata*) has a reddish brown center; in age, the cuticle outside the center breaks up into concentrically arranged scales, exposing the whitish background. Unlike *L. magnispora*, it has a smooth stalk. *Lepiota castaneidisca* is common under Coast Redwoods. **Lepiota sequoiarum** has a silky white cap 2 to 4 cm broad and a white, smooth, ringed stalk.

Lepiota rubrotinctoides

CAP: 2 to 8 cm in diameter, convex becoming nearly plane and broadly umbonate, the center brown to reddish brown, fading toward margin, the cuticle splitting radially over the white background, not changing color when bruised. **GILLS:** Free, close, white. **STALK:** 4 to 12 cm long, 3 to 7 mm wide, equal or enlarged at base, white, smooth, the partial veil leaving a white, persistent membranous ring. **SPORES:** White, elliptical, smooth, dextrinoid. **HABITAT:** Single or scattered on the ground in mixed hard-

Lepiota rubrotinctoides

wood/coniferous forests. **EDIBILITY:** Unknown; some small relatives of this species are dangerously poisonous.

This is a common mushroom of the forest soon after the first fall rains. The reddish brown cuticle splits as the cap expands, resulting in a radially fibrillose surface with dense colors in the center and lighter colors toward the margin. This species can be confused with **Lepiota castaneidisca,** but the cap of that species has a cuticle that breaks into concentric scales rather than splitting from the center of the cap to the margin. Other species with a "bulls-eye cap" include **L. flammeatincta,** which develops red stains on the cap and stalk, but not the gills, when bruised, and **Leucoagaricus erythrophaeus** (*Lepiota roseifolia*) which bruises red on all parts of the fruiting body.

Leucoagaricus leucothites

(Other name: *Lepiota naucina*)

CAP: 4 to 15 cm in diameter, hemispherical becoming convex to almost plane in age, white or grayish white, smooth, occasionally slightly scaly in dry weather, odor and taste mild. **GILLS:** Free, close, white developing pink tones in age. **STALK:** 5 to 12 cm long, 1 to 2 cm wide, often with a swollen base, white, dry, smooth, sometimes bruising yellow to brown, the partial veil leaving a persistent, membranous ring. **SPORES:** White, ellipti-

Leucoagaricus leucothites

cal, smooth, dextrinoid. **HABITAT:** Single or scattered in grassy areas in parks and gardens, and in disturbed soil, such as paths and roadsides. **EDIBILITY:** Edible but causes gastrointestinal upsets in some people. Caution is advised because it could be mistaken for the dangerously poisonous *Amanita ocreata*, which is also completely white.

The white cap and stalk, free gills, membranous ring, white spores, and grass habitat are the primary characteristics of this species. Although **Amanita ocreata,** the Destroying Angel (p. 74), has a volva at the base of the stalk and grows in mycorrhizal association with trees, it sometimes fruits in grassy areas some distance from the nearest tree. To be on the safe side, *L. leucothites* should not be collected for the table since confusion between the two species could have disastrous consequences.

Leucocoprinus birnbaumii
Flower Pot Mushroom
CAP: 2 to 6 cm in diameter, oval when young becoming bell shaped, bright yellow but fading to pale yellow, the margin distinctly striate, the surface dry, covered with powdery scales, flesh soft and yellow, not changing color when bruised. **GILLS:** Free, close, yellow to pale yellow. **STALK:** 3 to 8 cm long, 2 to 8 mm wide, equal or tapered toward the cap, yellow fading pale

Leucocoprinus birnbaumii

yellow, smooth or powdery, partial veil leaving a fragile yellow ring that may disappear. **SPORES:** White, elliptical, smooth, dextrinoid. **HABITAT:** Single, scattered, or in small clusters in flower pots and greenhouses or less often outside during summer in gardens, wood chips, and disturbed soils, such as along paths and roads. **EDIBILITY:** Unknown, possibly poisonous.

This eye-catching mushroom is quickly identified by its bright yellow color and tendency to fruit in indoor flower pots. It consumes organic matter in the potting mix as a saprobe and causes no harm to the plant, although the mycelium in the pot may become so dense that it is difficult to water the plant. The cap is striate and powdery or finely scaly. The related **L. cepistipes** (*Lepiota cepistipes*) has a white, scaly, striate soft cap. It grows in clusters on wood chips and lignin-rich soil during the warmer times of the year.

Lactarius Species

Milk caps (*Lactarius* species) are medium- to large-sized mycorrhizal partners of many trees. They have white to yellowish spores and firm flesh that, similar to *Russula* species, breaks cleanly (like a piece of chalk) due to the presence of clusters of spherical cells. In contrast, other mushrooms are typically fibrous due to elongated cells. The key feature of the milk caps, however, is the milklike latex that exudes from all parts of an injured fruiting body. In some species, the latex changes color when exposed to the air; in others, the latex stains tissue various colors. The latex may flow when the tissue is damaged, but in old specimens or in dry weather the latex may be scant. The taste of milk caps, which ranges from mild to extremely acrid (peppery), is used to identify the scores of species that occur in the West.

a. Latex orange...**Lactarius deliciosus**

a. Latex red................................**L. rubrilacteus** (see *L. deliciosus*)

a. Latex white or watery white, at least initially
    b. Latex quickly yellowing after exposure to air (within 30 seconds)
        c. Cap brownish orange, margin smooth...............................
        ....**L. xanthogalactus**

       c. Cap white, margin hairy...**L. resimus**
   b. Latex changing to yellow after a long delay, to a color other than yellow, or unchanging (but may discolor tissue)
       d. Fruiting body cinnamon to rusty brown; latex watery white; odor of maple syrup .... **L. rubidus**
       d. Fruiting body lacking reddish hues
          e. Stalk distinctly scrobiculate (pitted); cap yellowish .....
          .... **L. alnicola**
          e. Stalk not or faintly scrobiculate
             f. Cap flesh bruising lilac; latex staining wounded tissue lilac; cap gray to grayish brown......**L. pallescens**
             f. Cap flesh neither colored nor bruising lilac
                g. Cap pale grayish orange, tacky to viscid but not slimy ....**L. argillaceifolius**
                g. Cap brownish, viscid to slimy ....
                ....**L. pseudomucidus**

Lactarius alnicola
Golden Milk Cap

CAP: 6 to 14 cm in diameter, becoming plane or centrally depressed in age, the margin incurved and pubescent at first, smooth in age, the surface viscid when moist, various shades of yellow to ochre, the colors arranged in concentric bands (zonate) or not, flesh white, odor mild, taste very acrid. **LATEX:** White, slowly drying pale yellow. **GILLS:** Adnate to slightly decurrent, white becoming yellowish. **STALK:** 2 to 5 cm long, 1 to 2.5 cm wide, equal or tapered toward the base, colored like the cap, scrobiculate (decorated with darker yellow or ochre, round or irregular spots), partial veil absent. **SPORES:** White to pale yellow, elliptical, ornamented with warts and ridges. **HABITAT:** Single or gregarious in woods and landscaped areas under a variety of trees, especially common under oak. **EDIBILITY:** Inedible; the peppery taste is a deterrent.

The yellowish cap, white latex that slowly dries yellow, spotted stalk, and very acrid taste distinguish this common milk cap. It is encountered in many coastal and montane forests under a variety of trees and bushes. *Lactarius scrobiculatus* is similar but is readily distinguished from *L. alnicola* by a white latex that quickly turns yellow when exposed to air. It has a smooth or hairy cap margin and a slightly acrid taste.

Lactarius alnicola

Lactarius argillaceifolius

CAP: 8 to 20 cm in diameter, convex to plane, the center often depressed, grayish orange, grayish lilac, or light brown, azonate or faintly zonate near the margin, tacky or viscid when moist,

Lactarius argillaceifolius

smooth, flesh white to pale yellow, odor mild, taste mild to acrid. **LATEX:** Off-white or creamy, unchanging when exposed to air but staining tissue brownish. **GILLS:** Adnate to slightly decurrent, yellowish when young becoming yellow-brown in age and stained brown by the latex. **STALK:** 6 to 15 cm long, 2 to 5 cm wide, equal or slightly clavate, pale yellow or buff, partial veil absent. **SPORES:** Pale yellow, elliptical, ornamented with warts and ridges. **HABITAT:** Single or scattered in mycorrhizal association with oak in coastal and montane forests. **EDIBILITY:** Unknown.

The cap color of this robust milk cap is quite variable, but the colors are generally dull. A lilac tinge may or may not be present. The latex is cream colored, and stains gill tissue brown (but typically slowly). It might be confused with **L. pallescens** (p. 95), but that species is lighter in color and has a latex that stains gill tissue purplish, not brown. The edibility of these brownish or purplish staining milk caps is unknown.

Lactarius deliciosus

CAP: 4 to 12 cm in diameter, convex becoming plane, often with a depressed center in age, orange of various shades, often concentrically zoned, staining green in blotches or entirely green in age, surface viscid when moist, smooth, flesh orange, taste mild

Lactarius deliciosus

or slightly bitter. **LATEX:** Orange, slowly staining wounded tissue green. **GILLS:** Adnate to slightly decurrent, orange, becoming green when bruised and in age. **STALK:** 3 to 6 cm long, 1 to 2 cm wide, equal, orange, dry, staining green when bruised and in age. **SPORES:** Pale yellow, elliptical, ornamented with warts and ridges. **HABITAT:** Single, scattered, or gregarious under pines in coastal and montane forests. **EDIBILITY:** Edible, but many do not think its texture and taste live up to its name.

This species is easily identified by the orange cap, stalk, and gills that soon become partially or completely green. The name *L. deliciosus* was first used to describe a European species and was subsequently adopted in the United States; however, the true identity of our species is no doubt waiting to be described, a common theme when European names are used for U.S. specimens. In addition, *L. deliciosus* represents a complex of species. **Lactarius rubrilacteus** is very similar but has red latex. **Lactarius olympianus,** another orange milk cap, has white latex and an acrid taste. It is common in the Pacific Northwest.

Lactarius pallescens

CAP: 3 to 10 cm in diameter, convex to plane, often with a shallowly depressed center, gray to grayish brown, often with dull orange blotches, azonate or sometimes faintly zoned, smooth, very viscid when moist, flesh whitish, becoming lilac when bruised, odor mild, taste mild or faintly acrid. **LATEX:** White, drying lilac-brown, staining wounded tissue lilac. **GILLS:** Adnate to slightly decurrent, white to pale yellow, staining lilac when cut. **STALK:** 3 to 8 cm long, 1 to 2 cm wide, more-or-less equal, whitish or dingy yellowish, smooth, viscid when moist, staining lilac, solid, partial veil absent. **SPORES:** Pale yellow to pale orange, elliptical, ornamented with warts and ridges. **HABITAT:** Single or scattered in coastal and montane mixed hardwood/ coniferous forests. **EDIBILITY:** Unknown.

The variable color of the cap (grayish to brown, often with orange tones), viscid cap and stalk, and white latex that stains the gills lilac are the main field marks. Although old specimens may have a scant amount of latex, this species is still identifiable by the lilac stains. **Lactarius uvidus** is a similar purple-staining milk cap, but it has a pale lilac to lilac-gray cap, in addition to microscopic differences in the structure of the cap cuticle and

Lactarius pallescens

ornamentation of the spores, and different reactions to potassium hydroxide on the cap cuticle (yellow in *L. pallescens* and green in *L. uvidus*).

Lactarius pseudomucidus

CAP: 2 to 10 cm in diameter, convex to plane, often with a shallowly depressed center, brown to grayish brown but fading in age, smooth, azonate but may have water spots, margin inrolled at first, viscid to slimy, flesh grayish, not staining when bruised or cut, odor mild, taste slowly acrid. **LATEX:** White, unchanging when exposed but sometimes staining gills brown. **GILLS:** Adnate to slightly decurrent, whitish to pale orange, staining brown in age. **STALK:** 4 to 10 cm long, 0.5 to 1 cm wide, equal, colored like cap or slightly paler, smooth, viscid when moist. **SPORES:** White, elliptical, ornamented with warts and ridges. **HABITAT:** Scattered to gregarious under conifers in coastal forests. **EDIBILITY:** Unknown.

This species is apparently limited to forests of northern California and the Pacific Northwest, where it is common. It can be recognized by a pale grayish brown, viscid cap, pallid gills, and a stalk colored like the cap. When moist the cap and stalk are very slimy. *Lactarius kauffmanii* is a similar but larger milk cap of coastal forests of the West. It has a dark brown viscid cap and

Lactarius pseudomucidus

white latex that dries grayish green. The viscid stalk is usually paler brown than the cap. ***Lactarius argillaceifolius*** (p. 93) has a lighter grayish orange cap and is associated with oaks.

Lactarius resimus

CAP: 4 to 15 cm in diameter, convex but soon plane with a depressed center, the margin inrolled and bearded (covered with white wooly hairs), the surface strongly fibrillose becoming smooth in age, white, developing orange stains, viscid when moist, azonate, flesh white, yellowing when cut and exposed, odor mild, taste mild to slowly acrid. **LATEX:** White, quickly yellowing when exposed. **GILLS:** Adnate to slightly decurrent, close, white to pale yellow. **STALK:** 2 to 6 cm long, 1 to 3 cm wide, equal or tapered toward the base, white or tinged yellow, smooth but sometimes faintly scrobiculate (with pits and dots) in age, partial veil absent. **SPORES:** White to pale yellow, broadly elliptical, ornamented with warts and ridges. **HABITAT:** Scattered to gregarious on ground in hardwood/coniferous forests. **EDIBILITY:** Unknown.

Lactarius resimus is an example of a bearded milk cap. It is characterized by a white cap, white latex that quickly changes to yellow, and white stalk. ***Lactarius pubescens* var. *betulae*** is a similarly colored bearded milk cap, but it has a distinctly scro-

Lactarius resimus

biculate stalk (decorated with round or irregular pits and dots). It is limited to a mycorrhizal association with ornamental birch. **Lactarius torminosus,** found in montane forests with poplars and willows, is a cinnamon-colored bearded milk cap.

Lactarius rubidus

(Other name: *Lactarius fragilis* var. *rubidus*) Candy Cap

CAP: 2 to 5 cm in diameter, convex to plane, often with a shallowly depressed center, cinnamon brown or rusty brown, azonate, smooth to slightly roughened, dry or moist, flesh pale orange, taste mild, odor of maple syrup or butterscotch, especially upon drying. **LATEX:** Watery white, unchanging when exposed to the air. **GILLS:** Adnate to slightly decurrent, pale orange. **STALK:** 2 to 7 cm long, 0.5 to 1 cm wide, more-or-less equal, colored like cap, smooth, hollow in age. **SPORES:** White to pale yellow, subglobose, ornamented with warts and ridges. **HABITAT:** Single or scattered to densely gregarious, especially common under oaks in mixed hardwood/coniferous forests. **EDIBILITY:** Edible and good, sometimes used as a sweetener.

The Candy Cap is characterized by a cinnamon brown azonate cap, watery latex that does not change color, and sweet odor that may be faint in fresh specimens but intensifies as the fruit-

Lactarius rubidus

ing body dries. It often fruits in sizable numbers and is easily collected in bulk. Similarly colored milk caps include **L. rufulus**, which is larger than *L. rubidus*. It has a mild to faintly sweet odor, mild taste, solid or nearly solid stalk, and a preference for oaks; **L. rufus**, which has a unchanging white latex, distinctly acrid taste, and a preference for pines; and **L. subflammeus** with a bright orange, somewhat viscid cap, unchanging white latex, and a slowly developing, slightly acrid taste.

Lactarius xanthogalactus
(Other names: *Lactarius chrysorheus*, *L. vinaceorufescens*)
CAP: 3 to 10 cm in diameter, convex to plane, often centrally depressed, brownish orange, pinkish brown, or in age reddish brown, smooth, azonate or faintly zonate, often with darker spots scattered over the surface, odor mild, taste mild or slightly acrid. **LATEX:** White but quickly turning yellow when exposed to air. **GILLS:** Adnate to slightly decurrent, pale orange. **STALK:** 2 to 6 cm long, 1 to 2 cm wide, more-or-less equal, colored like cap or a little paler, smooth. **SPORES:** Pale yellow, elliptical, ornamented with warts and ridges. **HABITAT:** Scattered to gregarious under

Lactarius xanthogalactus

conifers and hardwoods in coastal and montane forests. **EDIBIL-ITY:** Unknown, but the acrid taste is a deterrent.

This common milk cap has a brownish orange cap and white latex that quickly changes to yellow when exposed to air (usually within seconds). But like all milk caps, the latex may be scant or nonexistent in old dry specimens. The stalk lacks pitted spots, or scrobiculations, common in many milk caps. In the field, this mushroom might be mistaken for **L. rubidus** (p. 98), but that mushroom is redder in color and has watery latex that does not change color, and a sweet odor. **Lactarius substriatus** has a reddish orange viscid cap, white latex that slowly turns yellow, smooth stalk, and a slightly acrid taste.

Russula Species

Like *Lactarius* (p. 91), the flesh of *Russula* breaks cleanly and audibly, but unlike *Lactarius*, latex is absent from the tissue. The West has probably well over 100 species, and many are very common. Identification of many species is extremely difficult, especially since most are not identical to their European counterparts and namesakes. An individual specimen can undergo a bewildering change in color as it matures. Accurate identification of many species requires examination of the ends of hyphal tips on the cap surface and a microscopic description of the spore ornamentation. A few are good edibles; others cause quick gastrointestinal upsets. All are mycorrhizal.

a. Gills of varying lengths; stalk very firm and rigid
   b. Gills not changing color when bruised.....**Russula brevipe**s
   b. Gills changing to red and then black when bruised..............
   .... **R. nigricans**

a. Gills usually of one length; stalk rigid but not usually firm
   c. Spores white
      d. Taste mild
         e. Cap brown, becoming greenish brown........................
         ....**R. brunneola**
         e. Cap lilac, becoming a mixture of colors in age.............
         .... **R. cyanoxantha**
      d. Taste acrid
         f. Cap pink or red .... **R. cremoricolor** (red phase)
         f. Cap white or yellow
            g. Cap white to pale yellow...........................**R. raoultii**
            g. Cap predominately cream colored ....
            .... **R. cremoricolor**
   c. Spores yellow
      i. Taste mild
         j. Cap olive green or brownish becoming vinaceous or purplish brown in age
            k. Cap olive green, vinaceous in age, robust, firm........
            .... **R. olivacea**
            k. Cap brown, aging pale purplish brown, fragile in age.. **R. integra**
         j. Cap predominately pink, red, or purple
            l. Cap purple, stalk white ....**R. amethystina**
            l. Cap red, burgundy, vinaceous, or pink
               m. Cap red, burgundy, or purplish brown; odor of shrimp; flesh slowly bruising brown ....
               ....**R. xerampelina**
               m. Cap pink, red, or variegated; odor mild; stalk white, often with yellow stains ....**R. abietina**
      i. Taste acrid
         n. Cap brownish

  o. Cap yellow-brown; odor strongly of almonds, fetid in age ....**R. fragrantissima**

  o. Cap brown, sticky, conspicuously striate; odor unpleasant .... **R. amoenolens**

 n. Cap reddish or variegated

  p. Cap and stalk red; taste immediately acrid ....
....**R. sanguinea**

  p. Cap reddish but highly variable; taste slowly very acrid.. **R. tenuiceps**

  p. Cap variegated, a mixture of purple, red, and olive, usually darker in center; odor of geraniums.............
....**R. pelargonia**

Russula abietina

CAP: 3 to 7 cm in diameter, convex to plane, at first pink to vinaceous with a darker reddish center, soon becoming variegated (a mixture of pink, purple, olive, and sometimes brown), the surface smooth, viscid when moist, the cap fragile in age, pileocystidia present, odor and taste mild. **GILLS:** Adnexed to adnate, white at first becoming yellow. **STALK:** 3 to 6 cm long, 1 to 2 cm

Russula abietina

wide, equal or clavate, white, often with yellow stains, especially on the base. **SPORES:** Yellow, subglobose, ornamented with warts less than 0.8 μm high with few connecting ridges. **HABITAT:** Gregarious in mixed hardwood/coniferous forests, at times very common. **EDIBILITY:** Unknown but too small and insubstantial to be of value.

The yellow spores, mild taste, pink or variegated cap, and especially the yellow stains on the stalk of this modestly sized *Russula* help identify it. ***Russula cessans*** is similar, but it has a rosy purplish cap that fades toward the margin in age and does not become variegated. It also has a mild taste. ***Russula queletii***, which is common under oaks has a purplish cap that fades in age, distinctive purple-tinged stalk, and an acrid taste.

Russula amethystina

CAP: 3 to 10 cm in diameter, convex to plane, light purple, becoming vinaceous or reddish in age, a faint glistening bloom is often visible on the surface of young specimens, cap viscid when moist, smooth, the margin with short striations in age, walls of hyphal tips on cap surface encrusted, taste mild. **GILLS:** Adnexed and adnate, yellow. **STALK:** 3 to 6 cm long, 1 to 2 cm wide, equal or slightly clavate, white. **SPORES:** Yellow, subglobose, orna-

Russula amethystina

mented with warts 0.2 to 0.7 µm high, forming a partial reticulum. **HABITAT:** Single or in small groups in mixed hardwood/coniferous forests; common but never in great numbers. **EDIBILITY:** Unknown.

In the field, a positive identification is difficult because many *Russula* species have purplish caps. In the laboratory, a quick look at the cap surface under the microscope reveals the crystalline bodies on hyphal tips, which narrow the identification to this species and its relatives. One is *R. turci,* a very similar species that may have an iodine-like odor in the base of the stalk; otherwise, it is indistinguishable. Another is *R. lilacea,* a small, fragile mushroom with a light purple cap with a dark purple center and very pale yellow spores. *Russula murrillii* has a deep purple cap without encrustations on free hyphal tips.

Russula amoenolens

CAP: 3 to 8 cm in diameter, convex at first becoming flat to centrally depressed, tan to yellowish brown to brown, sometimes fading toward the margin, viscid and sticky when moist, the margin often wavy in age, becoming deeply striate, pileocystidia

Russula amoenolens

absent, odor unpleasant, like burnt plastic, taste slowly acrid, unpleasant. **GILLS:** Adnexed to adnate, white aging brownish. **STALK:** 3 to 8 cm long, 1 to 2 cm wide, equal or slightly clavate, white, sometimes with brown stains. **SPORES:** Cream to pale yellow, subglobose, ornamented with warts 0.4 to 0.8 μm high and a few connecting ridges. **HABITAT:** Single or scattered under conifers and hardwoods. **EDIBILITY:** Inedible.

This common yellow-spored *Russula* is fairly easy to identify. The brown viscid cap is clearly striate in age, and it has an unpleasant odor and taste. It could be confused with **R. fragrantissima** (p. 109), but that species has a much larger yellow-brown cap with a distinctive odor of almond paste. **Russula integra** (p. 110) has a brown to purplish brown cap that lacks deep striations on the margin, and it has a mild odor. The spores of *R. integra* are ornamented with isolated spines.

Russula brevipes

CAP: 7 to 30 cm in diameter, convex to plane but soon with a depressed center or becoming funnel shaped, the margin often wavy, surface white but frequently covered with litter, some-

Russula brevipes

times with brownish stains in age, smooth, usually dry, flesh firm and rigid, unchanging when bruised or cut, pileocystidia absent, odor mild, taste mild to slowly acrid. **GILLS:** Adnexed to slightly decurrent, of various lengths, white. **STALK:** 3 to 9 cm long, 3 to 6 cm wide, more-or-less equal, white, sometimes with yellow or brown stains, smooth, firm, unchanging when bruised. **SPORES:** White to cream, subglobose, warts generally less than 1 µm high, partially reticulate. **HABITAT:** Single or scattered in mycorrhizal association with trees in coastal and montane forests, very common. **EDIBILITY:** Edible but of poor quality.

This common species is actually a complex of at least four species in the West. The white funnel-shaped cap often barely makes it through the duff on the forest floor as it pushes up needles and other leaves. Sometimes, it is identified by a telltale shrump—an elevated, roundish lump in the duff. One species in this complex has a fleeting light green color to the gills. Another species, which is similar to *R. chloroides* of Europe, has a greenish tinge to the gills and top of the stalk, but that coloration may disappear, making field identification difficult.

Russula brunneola

CAP: 6 to 12 cm in diameter, broadly convex and often slightly depressed in the center, uniformly brown or often somewhat mottled brown on a lighter brown background, the center remaining dark brown, finally greenish brown in age, the margin usually striate, pileocystidia absent, flesh unchanging when bruised, brittle, taste mild. **GILLS:** Adnexed to adnate, of one length, sometimes forked near the stalk, white. **STALK:** 5 to 12 cm long, 2 to 3 cm wide, dry, white but sometimes with a purplish tint in age. **SPORES:** White, subglobose, ornamented with warts less than 0.5 µm high, partially reticulate. **HABITAT:** Single or scattered in coastal and montane forests. **EDIBILITY:** Unknown.

The white spores and mild taste are not characteristics found in many brown *Russula* species in the western United States. Other species with white spores and a mild taste include the very common *R. cyanoxantha* (p. 108), which has a lilac or variegated cap, and at least two unnamed species: a species with a yellow, very firm cap with a mattelike finish, and a species with a white cap, white- to cream-colored spores, and a preference for oaks at low elevations.

Russula brunneola

Russula cremoricolor
(Other name: *Russula silvicola*)

CAP: 3 to 8 cm in diameter, convex to plane or shallowly depressed in age, cream colored, pink, or bright red, smooth, viscid when moist, the margin striate at maturity, flesh unchanging when bruised, pileocystidia abundant, taste quickly and persistently acrid. **GILLS:** Adnexed to adnate, white. **STALK:** 3 to 8 cm

Russula cremoricolor

long, 1 to 2 cm wide, equal or clavate, smooth, white. **SPORES:** White, subglobose, ornamented with warts less than 0.5 μm high with few connecting ridges. **HABITAT:** Scattered to gregarious in coastal and montane forests, often abundant during the middle of the mushroom season. **EDIBILITY:** Toxic, causing quick gastrointestinal upsets.

This species often fruits abundantly in many forests in the western United States. Based on DNA evidence, the red form, locally known as *R. silvicola* or *R. emetica*, is a color variant of the cream-colored form, named *R. cremoricolor*. Because *R. emetica* is a distinct mushroom and the name *R. cremoricolor* was used before the name *R. silvicola* was coined, the name *R. cremoricolor* has precedence. Obviously, the specific name *cremoricolor*, meaning "cream colored," was not meant for a bright red mushroom. The pink phase is not as common as the yellow or red variants, but all three mingle together, compounding the difficulties in field recognition. Even small amounts of this species will cause an upset stomach.

Russula cyanoxantha

CAP: 6 to 18 cm in diameter, convex but often slightly depressed in the center, uniformly lilac at first, becoming a mixture of lilac, olive, and yellow (variegated), smooth, dry, pileocystidia pres-

Russula cyanoxantha

ent, flesh unchanging when bruised, taste mild. **GILLS:** Adnexed, white and somewhat soft and flexible. **STALK:** 6 to 12 cm long, 2 to 5 cm wide, equal, white or white tinged with pink or lilac. **SPORES:** White, subglobose, warts mostly less than 0.4 μm high with few connecting ridges. **HABITAT:** Single or scattered on ground in forests, common. **EDIBILITY:** Edible and good.

This fairly robust species is common in both coastal and montane forests. The white spores, mild taste, and lilac color of the cap that becomes variegated in age are good field marks. Another distinguishing feature is the soft and flexible gill texture, which is unusual for *Russula* species. *Russula cyanoxantha* could be mistaken for another robust species, **R. olivacea** (p. 112), which may also develop a variegated cap, but that species has yellow, not white, spores. Many other *Russula* species have lilac or purplish caps, but those species have yellow spores, an acrid taste, or both, and most are smaller.

Russula fragrantissima

CAP: 6 to 20 cm in diameter, convex to plane, sometimes shallowly depressed in the center, yellow-brown with darker stains, the margin conspicuously striate in age, pileocystidia present, odor strongly of almond paste but fetid in older specimens, taste

Russula fragrantissima

bitter and then acrid. **GILLS:** Adnexed and adnate, white when young, tan in age with reddish brown stains, often beads of moisture visible on the edge of the gills. **STALK:** 7 to 15 cm long, 1 to 4 cm wide, more-or-less equal, white or light tan, often with brownish stains. **SPORES:** Pale yellow, subglobose, ornamented with warts up to 1.2 μm high, forming a nearly complete reticulum. **HABITAT:** Often singly but sometimes scattered or even in troops in mixed hardwood/coniferous coastal and montane forests throughout the West. **EDIBILITY:** Inedible.

The odor of almond paste is usually so distinctive that this mushroom is easily identified in the field. It has a relatively large yellow-brown cap with a striate margin, especially in age. Both microscopic and macroscopic characteristics describe at least two species found in the forests of the West. Neither is probably the true *R. fragrantissima*, but they both have clear affinities to that species. None is edible due to the disagreeable taste.

Russula integra

CAP: 6 to 12 cm in diameter, convex to plane or slightly centrally depressed, dark brown but soon becoming purplish brown, pinkish brown, or light brown, the center usually, but not always, remaining dark brown, the margin smooth or slightly striate, the surface dry, with few encrusted hyphal tips and pileocystidia, flesh fragile, taste mild. **GILLS:** Adnexed to adnate, white but soon yellow and then deep yellow. **STALK:** 5 to 12 cm long, 2 to 3 cm wide, equal or clavate, white, fragile in age. **SPORES:** Dark yellow, subglobose, ornamentation almost exclusively isolated spines up to 2 μm high. **HABITAT:** Single or scattered in mixed hardwood/coniferous forests, common and one of the first *Russula* species of the mushroom season. **EDIBILITY:** Unknown, but often very fragile in age and not a likely candidate for the table.

This is a brown-capped species with a white stalk, yellow spores, and mild taste. Because it goes through so many color changes and does so in rapid succession (even for a *Russula*), it is difficult to call it any one color. In most specimens, there remains some hint of brown, but often other colors, especially purple, dominate. **Russula mustelina** has a brown cap, white stalk, pale yellow spores, and very firm flesh, an unusual feature for members of this group of *Russula* species.

Russula integra

Russula nigricans

CAP: 6 to 20 cm in diameter, convex becoming flat to centrally depressed in age, white but aging dull gray to black, smooth, dry, pileocystidia absent, flesh white bruising red and then black, firm, odor mild, taste mild to slightly acrid. **GILLS:** Ad-

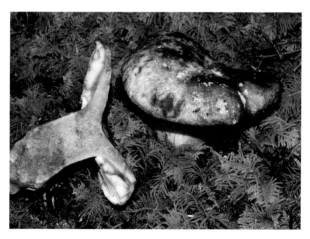

Russula nigricans

nexed to slightly decurrent, of various lengths, subdistant to distant, thick, brittle, white, bruising red and then black. **STALK:** 5 to 8 cm long, 2 to 5 cm wide, more-or-less equal, white becoming gray to black in age, flesh white, bruising like cap and gills. **SPORES:** White, subglobose, ornamented with warts to 0.3 μm high, partially reticulate. **HABITAT:** Single or scattered in forests, common but easily overlooked since older caps are blackish and sometimes covered with duff. **EDIBILITY:** Unknown.

The unambiguous color change from white to red to black of all parts of this mushroom is distinctive. The development of the red color occurs in seconds to a few minutes, whereas the change to black may take many minutes to develop. Similar species with a red-then-black bruising reaction include *R. acrifolia,* which has a very acrid taste, and *R. dissimulans,* which has close to subdistant gills and larger spores than *R. nigricans.* **Russula albonigra** is distinctive because it bruises directly from white to black (no red interphase).

Russula olivacea

CAP: 10 to 30 cm in diameter, convex to plane, olive green at first, becoming variegated with reddish colors or completely vinaceous in age, dull, dry, smooth, firm, the cuticle separating

Russula olivacea

from the cap only at the margin, pileocystidia absent, odor and taste mild. **GILLS:** Adnexed to adnate, white but soon pale yellow. **STALK:** 7 to 18 cm long, 2 to 6 cm wide, equal to clavate, white with a pink flush over some or all of the stalk, dry, smooth. **SPORES:** Yellow, subglobose, ornamented with isolated spines up to 2 μm high. **HABITAT:** Single or scattered in coastal and montane forests, favoring conifers. **EDIBILITY:** Recommended by some but apparently causing gastrointestinal upsets in others.

The olive color of the cap, at least when the fruiting body is young, firm texture, robust size, yellow spores, and mild taste are the field marks of this common mushroom. Specimens as large as dinner plates are not uncommon. In age, the cap may become partially or completely vinaceous, but the dry dull texture of the cap and pink blushes of the stalk help identify older specimens. In the western United States, this is actually a complex of at least three species. All need names since the true *R. olivacea* of Europe is probably distinct from any of ours. The edibility of all the different types in the United States is unknown.

Russula pelargonia

CAP: 3 to 7 cm in diameter, convex to plane, vinaceous but soon variegated, a mixture of purple, red, and olive with a darker center, *or* rarely completely pale yellow, viscid when moist, smooth, pileocystidia present, flesh unchanging when bruised, odor distinctly geranium-like but sometimes fruity in older specimens, taste acrid. **GILLS:** Adnexed to adnate, white aging yellow. **STALK:** 3 to 6 cm long, 1 to 1.5 cm wide, more-or-less equal, smooth, white, with tinges of rose or purple in age. **SPORES:** Yellow, subglobose, ornamented with warts 0.4 to 0.9 μm high, forming a partial reticulum. **HABITAT:** Scattered in coastal and montane forests. **EDIBILITY:** Unknown.

This is a fairly common mushroom of coastal and mountain forests. Like all *Russula* species, it grows in mycorrhizal association with various trees. It has yellow spores, an acrid taste, and two distinguishing features: a variegated cap and a pronounced geranium or fruity odor. Typical specimens are often olive and rose colored, with a darker reddish center. However, completely yellow specimens occur, causing no end of confusion for those who try to identify *Russula* species in the field. The cap of **R. bicolor,** common under spruce and pines along the coast and low

Russula pelargonia

to high elevations in mountains, is a mixture of pink and yellow. It also has an acrid taste but unlike *R. pelargonia*, it has white spores and no odor.

Russula raoultii

CAP: 3 to 8 cm in diameter, convex to plane, slightly depressed in age, white to pale yellow, sometimes with a faint gray tint in age, smooth, viscid, the margin smooth or faintly striate, pileocystidia present, flesh unchanging when bruised, taste acrid. **GILLS:** Adnexed to adnate, white. **STALK:** 3 to 8 cm long, 1 to 2 cm wide, equal or clavate, smooth, white or pale yellow. **SPORES:** White, subglobose, ornamented with warts up to 0.8 μm high, forming a nearly complete reticulum. **HABITAT:** Single or scattered in coastal and montane hardwood/coniferous forests. **EDIBILITY:** Inedible; the acrid taste is unpalatable for most.

White to pale yellow species of *Russula* with white spores and an acrid taste are difficult to distinguish from one another, with or without microscopic examination. This species tends to be white or very pale yellow, which helps to separate it from the very abundant *R. cremoricolor* (p. 107), which in the pale form is more cream colored than white (but may be pale yellow in

Russula raoultii

age) and is also more sharply and persistently acrid than is *R. raoultii*. **Russula crassotunicata** is similar in most respects, but the cuticle is thick and rubbery. **Russula stuntzii** has a white to grayish white cap with a slight tinge of lilac. It is unusual in that it often fruits on top of rotting logs.

Russula sanguinea
(Other name: *Russula rosacea*)

CAP: 4 to 12 cm in diameter, convex, plane, or shallowly depressed, scarlet red fading to pink in age, the margin usually striate in age, cuticle separating from cap with difficulty, pileocystidia present, odor mild, taste acrid. **GILLS:** Adnexed to adnate, white but soon becoming yellow. **STALK:** 5 to 10 cm long, 1 to 3 cm wide, more-or-less equal, pink or red, sometimes bright red. **SPORES:** Yellow, subglobose, ornamented with isolated warts 0.2 to 0.4 µm high. **HABITAT:** Single or scattered in mycorrhizal association with pines in coastal and montane forests. **EDIBILITY:** The acrid taste is a deterrent.

This attractive mushroom is common under pines in many habitats. The scarlet red cap and deep red or at least dark pink stalk, yellow spores, and acrid taste set it apart from other red

Russula sanguinea

members of *Russula*. **Russula rhodopoda,** which is similarly colored, has a mild but sometimes weakly acrid taste and yellow spores. It differs from *R. sanguinea* by its reticulate spores and the nature of the cuticle, which can be easily separated from the cap almost all the way to the center. In contrast, the cuticle of *R. sanguinea* separates with difficulty. **Russula queletii** has a purplish cap that fades in age, distinctive purple-tinged stalk, an acrid taste, and a preference for oaks.

Russula tenuiceps

CAP: 7 to 15 cm in diameter, convex to plane, sometimes with a shallowly depressed center, red or vinaceous but highly variable—often pink with a yellow center in age, but it may also become orange and red, purplish red, or even brownish, pileocystidia present, texture very fragile in age, odor mild, taste slowly very acrid. **GILLS:** Adnexed to adnate, white but soon yellow. **STALK:** 4 to 12 cm long, 1 to 3 cm wide, equal or clavate, dry, smooth, white or less often white with a pink blush, fragile. **SPORES:** Yellow, subglobose, ornamented with warts up to 1.2 μm high with few connecting ridges. **HABITAT:** Single or gregarious in mixed hardwood/coniferous forests. **EDIBILITY:** Inedible;

Russula tenuiceps

the acrid taste and fragile texture of all parts of the fruiting body are deterrents.

Russula tenuiceps is sometimes abundant in many of our western forests. The cap color is highly variable but is usually some shade of red. Yellow spores and the very fragile texture of all parts of this mushroom are good field characteristics, but the diagnostic feature is the *delay* (10 or more seconds) of the very acrid taste. The cap of **R. californiensis**, abundant under Monterey pine and not uncommon under Bishop pine, is brick red. It has a grayish staining stalk and mildly acrid taste.

Russula xerampelina
Shrimp Russula
CAP: 6 to 20 cm in diameter, convex becoming plane, color highly variable—burgundy to reddish purple is perhaps the most commonly encountered range of colors, but bright red, yellowish, and green caps also occur; the margin sometimes weakly striate in age, pileocystidia present, flesh white, bruising yellow and finally brown (this may take a few minutes), taste mild, odor of fish, shrimp, or crab, but this is not always appar-

ent to everyone. **GILLS:** Adnexed to adnate, white but soon yellow. **STALK:** 4 to 12 cm long, 2 to 4 cm wide, equal, white or white with a pink blush, smooth, bruising brown. **SPORES:** Yellow, subglobose, ornamented with warts up to 1 μm high with few connecting ridges. **HABITAT:** Single or scattered in coastal and montane forests. **EDIBILITY:** Edible.

Although the color of the cap is highly variable, the yellow spores, mild taste, fishy odor, and tendency of all parts of the fruiting body to stain yellow and finally brown help define this common mushroom. If in doubt, bruise the stalk and examine the bruised area after 20 minutes, which is ample time for the browning reaction to occur. This description represents a complex of five unnamed or difficult to distinguish species in the West. Other yellow-spored, mild-tasting species include ***R. basifurcata,*** an oak-loving species that has a cream-colored cap with orange blotches; it has forked gills and a mild odor.

Russula xerampelina

Waxy Caps

These are often brightly colored mushrooms with thick waxy gills and white spores. Members of the genus *Hygrocybe*, which are generally small in size, are apparently saprobic on leaf litter, whereas the members of *Hygrophorus*, which range in size from small to relatively large, are mycorrhizal. This group has few good edibles.

a. Cap green, yellow, orange, pink, or red, at least in part
  b. Cap green .... **Hygrocybe psittacina**
  b. Cap yellow or orange
    c. Cap conical, bruising black .... **H. conica**
    c. Cap conical, not bruising black ....
      .... **H. acutoconica** (see *H. flavescens*)
    c. Cap convex, not changing color when bruised ....
      .... **H. flavescens**
  b. Cap pinkish or red
    d. Cap bright red
      e. Cap viscid; stalk yellow and red .... **H. punicea**
      e. Cap not viscid, red; stalk mostly red .... **H. coccinea**
      e. Cap not viscid, red but fading to orange.......................
      .... **H. miniata** (see *H. coccinea*)
    d. Cap some shade of pink
      f. Cap pinkish orange, fruiting near snow........................
      .... **Hygrophorus goetzii**
      f. Cap pink and streaked vinaceous .... **H. russula**
      f. Cap and stalk uniformly pink .... **H. pudorinus**

a. Cap gray or white
  g. Cap gray; odor of almonds .... **H. agathosmus**
  g. Cap white; odor mild or cedarlike
    h. Cap and stalk viscid; odor mild..................... **H. eburneus**
    h. Cap and stalk moist but not viscid; odor of cedar ....
    .... **Hygrocybe russocoriacea**

Hygrocybe coccinea

CAP: 2 to 5 cm in diameter, conical to convex, sometimes with a small umbo, scarlet when fresh, fading in age to reddish orange, surface smooth, dry or moist but not viscid. **GILLS:** Adnexed to

Hygrocybe miniata

notched, waxy, reddish orange, fading paler, the edges often lighter than the sides. **STALK:** 2.5 to 5.0 cm long, 0.5 to 1 cm wide, smooth, hollow, dry to moist but not viscid, red or reddish orange but yellowish at the very base, partial veil absent. **SPORES:** White, elliptical, smooth, inamyloid. **HABITAT:** Scattered on ground in mixed hardwood/coniferous forests, especially in

Hygrocybe coccinea

colder protected canyons and drainages. **EDIBILITY:** Unknown but too small to be of value.

Hygrocybe coccinea is one of many colorful waxy caps that fruit in the middle of the mushroom season when temperatures are relatively cool. Other red-capped waxy caps include **H. puni-cea** (p. 124), which has a larger viscid cap (up to 10 cm or more in diameter) and a stalk that is usually yellow or a mixture of yellow and red (often with a whitish base) and longitudinally striate rather than smooth; **H. miniata** (p. 120), with a nonviscid red cap (1 to 3 cm broad) that fades to yellowish to orange and a red stalk; and **H. subminiata,** which has a small, viscid, scarlet cap (more than 1.5 cm broad) and dry yellow stalk.

Hygrocybe conica
Witch's Hat
CAP: 2 to 7 cm in diameter, distinctly conical, sometimes be-coming umbonate in age with an upturned margin, yellow, yel-lowish green, orange, or sometimes reddish, becoming black when bruised or in age, odor and taste mild. **GILLS:** Adnexed to free, close, soft and waxy, whitish to yellowish, eventually black-ening. **STALK:** 5 to 10 cm or more long, 0.5 to 1 cm wide, fragile,

Hygrocybe conica

equal, hollow, twisted, striate and easily splitting, yellow to orange with a whitish base, moist but not viscid, bruising and aging black, partial veil absent. **SPORES:** White, elliptical, smooth, inamyloid. **HABITAT:** Single or scattered in woods, especially common in Coast Redwood forests. **EDIBILITY:** Unknown.

This common waxy cap is distinctive for its yellow to red conical cap, waxy gills, and inclination to bruise and age black. Entirely black specimens are sometimes encountered. Like many of the brightly colored waxy caps, it fruits in cool weather. *Hygrocybe singeri* is similar, but the surface of both the cap and stalk is viscid, especially when moist. Another yellow, conical waxy cap, *H. acutoconica* (p. 4), does not bruise or age black. All of these mushrooms have waxy gills, white spores, and lack a partial veil.

Hygrocybe flavescens

CAP: 2.5 to 6 cm in diameter, convex to plane, yellow-orange when young, bright yellow when mature but sometimes yellowish orange in the center, smooth, viscid when moist, odor and taste mild. **GILLS:** Adnexed, soft, waxy, pale yellow to yellow. **STALK:** 4 to 7 cm long, 0.5 to 1.5 cm wide, equal, fragile and splitting easily, lubricious, yellow with a paler base, decorated with minute flecks or smooth, partial veil absent. **SPORES:** White, el-

Hygrocybe flavescens

liptical, smooth, inamyloid. **HABITAT:** Scattered or in groups on ground in woods. **EDIBILITY:** Unknown.

This widespread waxy cap is completely yellow or yellow-orange and does not change colors when bruised. It often grows in moist mossy places. *Hygrocybe acutoconica* (p. 4) is similarly colored, but its cap is clearly conical, and the stalk, which is usually striate, has a whitish base. Because the taxonomy of these and other waxy caps is unsettled, some of the names used here are placeholders until the species in the West are adequately described. *Hygrophorus chrysodon* is a whitish waxy cap with yellow granules on the cap and stalk. The granules may weather or wash off, but the apex of the stalk usually has a telltale sign of yellow. As a group, waxy caps are characterized by their soft, relatively thick, waxy gills and less often by their waxy-textured caps.

Hygrocybe psittacina
Parrot Waxy Cap
CAP: Up to 4 cm in diameter, bell shaped to convex to plane in age, bright, dark, or olive green but fading to olive orange and developing yellow, pink, and brown colors in age, viscid when moist, the margin translucent-striate, odor and taste mild. **GILLS:**

Hygrocybe psittacina

Adnate, yellowish green to light green. **STALK:** 2 to 8 cm long, 3 to 5 mm wide, viscid, equal, light green fading yellow or orange in age, hollow, partial veil absent. **SPORES:** White, elliptical, smooth, inamyloid. **HABITAT:** On the ground in the woods. **EDIBILITY:** Edible but too small to be of value for the table. There have been reports of antitumor properties when fed to laboratory mice.

This is an easy mushroom to identify. The small size, green color, and viscid translucent-striate cap (the striations are easily visible through the surface of the cap) are diagnostic. This is the only small, viscid, green mushroom likely to be encountered in the woods. Another attractive green waxy cap, **H. virescens,** is completely lime green or light green; the cap is moist but not viscid, and the margin is not striate. Like many waxy caps, it often grows in cool protected areas in the forest.

Hygrocybe punicea
Scarlet Waxy Cap

CAP: 3 to 12 cm in diameter, convex becoming flat in age, surface viscid when moist but dry and lubricious in dry weather, smooth, and shiny, bright red fading in streaks to reddish orange in age, texture waxy, flesh thin, fragile, and red or orange, odor and taste mild. **GILLS:** Adnexed, subdistant, red to reddish orange, fading to yellow, waxy. **STALK:** 4 to 14 cm long and up to 2 cm wide, equal or narrowing toward the base, yellow or a mixture of red and yellow, the base white or sometimes yellow, the surface dry, longitudinally streaked, partial veil absent. **SPORES:** White, elliptical, smooth, inamyloid. **HABITAT:** Single, scattered, or gregarious on the ground in hardwood/coniferous forests. **EDIBILITY:** Not recommended, apparently causing gastrointestinal upsets in some.

This striking waxy cap is one of several bright red species in western forests. *Hygrocybe punicea* has a bright red viscid cap and yellow and red stalk with longitudinal streaks; **H. coccinea** (p. 119) has a nonviscid red cap up to 5 cm broad and predominately red stalk that lacks streaks; **H. miniata** (p. 120) has a nonviscid red cap 1 to 3 cm broad that fades to yellowish to orange and a red stalk. All of these occur in cool damp places in forests.

Hygrocybe punicea

Hygrocybe russocoriacea

(Other name: *Camarophyllus russocoriaceus*)

CAP: 1 to 3 cm in diameter, convex, surface smooth, moist but not viscid, white or off-white, odor of cedar, taste mild. **GILLS:** Decurrent, distant, thick, waxy, white. **STALK:** 3 to 10 cm long, 3 to 5 mm thick, equal and often curved, colored like the cap, moist but not viscid, partial veil absent. **SPORES:** White, elliptical, smooth. **HABITAT:** Single or scattered in mixed woods along the central and northern California coast and Pacific Northwest, common but not fruiting in large numbers. **EDIBILITY:** Unknown but too small to be of value.

This small waxy cap is distinctive due to its odor of cedar, white moist cap, and long slender stalk. The species name means "Russian leather." If the odor isn't immediately noticeable, crush a piece of the stalk. Like most waxy caps, the gills are thick, distant (well spaced), and waxy. To detect the waxy texture, rub gill tissue between your fingers. Similar species include **H. virginea** (*Camarophyllus virgineus*), which has a whitish or yellow-stained cap up to 7 cm broad and mild odor. The relatively ro-

Hygrocybe russocoriacea

bust **H. pratensis** (*Camarophyllus pratensis*) has a dry, pale orange cap up to 8 cm broad and similarly colored gills and stalk
(up to 2 cm wide). It is found in mixed coastal forests and occasionally grassy areas.

Hygrophorus agathosmus

CAP: 3 to 10 cm in diameter, convex to nearly plane, gray to
brownish gray, often darker in the center, surface viscid when
moist, smooth, the margin incurved at first, flesh white, odor
fragrant, almondlike. **GILLS:** Adnate to slightly decurrent, waxy,
soft, white. **STALK:** 4 to 15 cm long, 1 to 2 cm wide, white or
tinged gray, more-or-less equal, nonviscid, smooth, partial veil
absent. **SPORES:** White, elliptical, smooth. **HABITAT:** Scattered to
gregarious under conifers. **EDIBILITY:** Edible but of poor quality.

This attractive waxy cap is characterized by a gray cap, white
gills and stalk, and almond odor. **Hygrophorus bakerensis** also
has a sweet almond odor. It has a reddish brown to yellow-
brown cap (4 to 15 cm broad) with a whitish margin, waxy
white gills, and white stalk. It is very common under conifers in
the Pacific Northwest. **Hygrophorus hypothejus** has a viscid,
variegated cap (brownish or olive brown in the center and
yellow-orange on the margin) up to 8 cm broad, pale yellow to

Hygrophorus agathosmus

orange, waxy, decurrent gills, viscid yellowish stalk, mild odor, and an association with pines. None of these species is considered a good edible.

Hygrophorus eburneus
Ivory Waxy Cap

CAP: 2 to 7 cm in diameter, convex to plane, the margin sometimes upturned in age, white, smooth, viscid to slimy in wet weather, developing yellow tints in age, unchanging when cut or bruised, odor and taste mild. **GILLS:** Adnate to decurrent, soft and waxy, white. **STALK:** 4 to 12 cm long, 0.5 to 1.5 cm wide, equal or tapered toward the base, white, viscid in moist weather, partial veil absent or leaving a faint layer of slime. **SPORES:** White, elliptical, smooth, inamyloid. **HABITAT:** Scattered, gregarious, or in clusters on ground in hardwood/coniferous forests. **EDIBILITY:** Edible but slimy. The extremely viscid cap attracts forest duff and may require attentive cleaning.

The pure white color, extremely viscid surface of the cap and stalk, and waxy gills are the distinguishing features of this common waxy cap, which has been called "Cowboy's Handkerchief" in reference to its color and slimy coating. Like many other waxy caps, it prefers cool temperatures. *Hygrophorus subalpinus* is common in the mountains under conifers near melting

Hygrophorus eburneus

snow in spring. It has a white viscid cap, white, stout, nonviscid stalk, and an evanescent ring. It often remains partially hidden under the duff. It is a common member of the snowbank flora in western mountains.

Hygrophorus goetzii

CAP: Up to 5 cm in diameter, convex becoming plane, pinkish orange, smooth, viscid, odor and taste mild. **GILLS:** Adnate, sub-

Hygrophorus goetzii

distant to distant, colored like cap or somewhat paler, soft, waxy. **STALK:** 3 to 6 cm or more long, 3 to 8 mm wide, colored like cap and lightly streaked, dry, smooth but with a hairy base, partial veil absent. **SPORES:** White, elliptical, smooth, inamyloid. **HABITAT:** Growing singly or in small groups under conifers in spring in the mountains, often close to melting snow. **EDIBILITY:** Unknown.

This attractive mushroom is a common member of the snowbank mushrooms—those mushrooms that follow the melting snow in spring and early summer in the mountains. Although other snowbank mushrooms fruit very close to the edge of melting snow, *H. goetzii* is distinctive because it sometimes grows through the snow. Like many of the waxy caps, this mushroom is a colorful and bright part of the forest flora. ***Hygrophorus subalpinus*** is another common snowbank mushroom. It is all white in color and has a rather stout stalk with an evanescent ring. It often remains partially hidden under the duff soon after snowmelt.

Hygrophorus pudorinus

CAP: 5 to 15 cm in diameter, convex becoming nearly flat in age, pink to pinkish orange, sometimes becoming bright reddish orange in age, smooth, viscid, the margin inrolled at first, flesh

Hygrophorus pudorinus

white to pink, odor mild. **GILLS:** Adnate to decurrent, soft and waxy, pale pink. **STALK:** 4 to 16 cm long, 1 to 3 cm wide, more-or-less equal, moist but not viscid, white to pink, upper part minutely scaly, partial veil absent. **SPORES:** White, elliptical, smooth. **HABITAT:** Scattered to gregarious, associated with conifers, especially common under spruce. **EDIBILITY:** Edible.

This large attractive waxy cap has a pink viscid cap and a moist, but not viscid, stalk. In age it develops bright orange tones. *Hygrophorus bakerensis* has a viscid cap that is brownish in the center and white on the margin, thick white gills, a non-viscid white stalk, and a distinct almond odor. *Hygrophorus agathosmus* (p. 126) has a viscid, smooth, pale gray cap up to 10 cm broad, white gills, nonviscid white stalk, and an almond odor. It grows under various conifers. The pink *H. goetzii* (p. 128) is smaller and grows in the high mountains near melting snow in spring.

Hygrophorus russula

CAP: 5 to 12 cm in diameter, convex to plane, pink with vinaceous or purplish red streaks and fibrils, viscid when moist, odor and taste mild. **GILLS:** Adnate to slightly decurrent, close or crowded, white to pink, often stained with vinaceous spots and streaks in age, soft and waxy. **STALK:** 2 to 10 cm long, up to 2.5

Hygrophorus russula

cm or more wide, colored and stained like cap, not viscid, equal, solid, partial veil absent. **SPORES:** White, elliptical, smooth. **HABITAT:** Single, scattered, or gregarious in hardwood/coniferous forests, often under oaks. **EDIBILITY:** Edible but of poor quality.

This attractive waxy cap is often fairly robust and resembles a vinaceous species of *Russula*. However, the tissue of *Hygrophorus* is fibrous, whereas the cap and stalk tissue of *Russula* breaks like chalk. In addition, the gills of *Hygrophorus* are waxy, not brittle like the gills of *Russula*. **Hygrophorus purpurascens** has a similarly colored cap (a background pink color overlaid with streaks of purplish red or vinaceous red), red spots on mature gills, and a partial veil that leaves an ephemeral fibrillose ring. It is common in the mountains under conifers soon after the snow melts and occasionally after summer rains.

Schizophyllum commune
Split Gill

CAP: 1 to 4 cm broad, fan shaped in general outline, typically shelflike, tough and leathery, surface dry, white to gray, covered with white or pale gray hairs, the margin inrolled and scalloped in dry weather, shriveling when dry and reviving in rainy weather, flesh thin, tough, and off-white in color. **GILLS:** Gray,

Schizophyllum commune

split lengthwise (thus appearing in pairs), distant. **STALK:** Absent. **SPORES:** White, cylindrical, smooth. **HABITAT:** Scattered, in rows, or in clusters on dead hardwood, causing a white rot. **EDIBILITY:** Inedible and, from some reports, to be avoided because it may swell when eaten due to its ability to revive when moist. It can remain dry for a half century and, with moisture, rehydrate. The mushroom has been implicated in a variety of adverse medical conditions, including sinusitis, allergies, and lung disease. On the other hand, some reports indicate that it has beneficial compounds with antitumor properties and is active against viruses.

Superficially resembling a polypore but related to the gilled mushrooms, the Split Gill is found on every continent except Antarctica. The split gills (appearing in pairs), hairy, whitish gray cap, and growth on hardwood twigs, branches, and logs are a unique combination of characteristics. It is encountered any time of the year and may be found on wooden construction materials.

Pleurotus ostreatus
Oyster Mushroom

CAP: 5 to 20 cm broad, typically fan shaped and growing laterally but occasionally circular and funnel shaped when growing vertically, white, gray, tan, or brown, smooth, margin inrolled when young, becoming wavy, flesh white, odor and taste mild. **GILLS:** Decurrent, white. **STALK:** Absent or if present, short, lateral or occasionally central, partial veil absent. **SPORES:** White, elliptical, smooth, inamyloid. **HABITAT:** In shelflike clusters or tiers or occasionally upright on fallen branches and logs of hardwood or, less often, conifer trees, or on shrubs or standing trees in forests and urban areas. **EDIBILITY:** Edible and choice.

Oyster Mushrooms usually grow shelflike on wood from a short lateral stalk and are easy to identify based on the above general description. However, a complex of similar species may exist. For example, the large meaty Oyster Mushrooms that grow on cottonwoods may be a different species than the ones found on oaks. The Oyster Mushroom is easily cultivated on almost any source of cellulose, including rice hulls, paper, coffee grounds, and corn cobs. Similar mushrooms include *Pleurocybella porrigens*, a white, small (cap 4 to 6 cm broad), thin, fan-

Pleurotus ostreatus

shaped species that grows on decaying conifers in the woods, and *Hohenbuehelia petaloides,* which has a shoehorn-shaped brown cap 3 to 6 cm broad. It is not uncommon in wood chips in landscaped areas. See *Phyllotopsis nidulans* (p. 192) for wood-inhabiting species with hairy caps and short or no stalks.

Cystoderma fallax

CAP: 2 to 4 cm in diameter, convex to plane, sometimes with a low umbo, cinnamon brown to yellow-brown, dry, coated with granules but less so in age, odor and taste mild. **GILLS:** Adnexed, close, white, cream, or buff. **STALK:** 2 to 6 cm long, 3 to 5 mm wide, equal, smooth or minutely fibrillose and whitish or pale yellow above the ring, coated with a conspicuous layer of brown or yellowish small granules or scales below the ring, partial veil leaving a flaring, membranous ring that is whitish on the upper surface and colored like the cap on the under surface, with veil remnants sometimes also adhering to the cap margin. **SPORES:** White, elliptical, smooth, and amyloid. **HABITAT:** Single, scattered, or in small clusters on ground, moss, and rotting logs in hardwood/coniferous forests. **EDIBILITY:** Unknown.

The cinnamon brown color, persistent membranous ring, white spores, and especially the granules that decorate the cap

Cystoderma fallax

and stalk are distinguishing features of this small mushroom. The granules may wash off the cap during rainy weather, but they usually adhere to the stalk. **Cystoderma amianthinum** is similar, but it has a fragile ring that may disappear. Both of these mushrooms seem to favor mossy areas.

Armillaria mellea
Honey Mushroom

CAP: 3 to 15 cm in diameter, color highly variable—tan, honey, pinkish tan, light brown, or gold, with circles of dark fibrils; flesh whitish to tan, odor mild, taste mild to bitter. **GILLS:** Attached but sometimes notched or slightly decurrent, off-white at first but discoloring pinkish to light brown in age. **STALK:** Up to 16 cm long, whitish becoming rusty brown, tough and fibrous, the partial veil leaving a ring (or traces of a ring) on upper stalk. **SPORES:** White, elliptical, smooth, inamyloid. **HABITAT:** In dense clusters (or less commonly, singly or scattered) on wood, often on stumps or diseased trees, but also fruiting on the ground from buried wood. **EDIBILITY:** Tasty and flavorful, but not agreeable to everyone. The stalks are tough and usually not eaten.

Armillaria mellea

The Honey Mushroom is a parasite of oaks and other trees, including domesticated plants. For example, it parasitizes and kills grapevines and fruit trees, causing significant economic losses to those crops. The mycelium is bioluminescent, and infected wood may glow in what is called "foxfire." Thick dark strands of mycelium, called rhizomorphs, grow through the soil and advance the fungus from one tree to another. Because of the sheer volume of mycelia and rhizomorphs in some forests, it has been claimed that some species of *Armillaria* are among the largest organisms on earth. **Armillaria gallica** fruits alone or in small groups, usually on the ground. The stalk base is typically swollen.

Floccularia luteovirens
(Other name: *Floccularia straminea*)

CAP: 4 to 15 cm in diameter, convex to plane, bright yellow but fading in age to pale yellow, the surface with conspicuous soft scales, flesh whitish, taste and odor mild. **GILLS:** Adnexed to notched, pale yellow, close. **STALK:** 4 to 12 cm long, 1 to 3 cm wide, equal or wider at the base, white and smooth near the apex, white and conspicuously scaly below the ring, the scales

Floccularia luteovirens

soft and edged yellow, often arranged in concentric bands around the stalk, partial veil leaving a fragile, white, cottony ring on the stalk and white remnants on the margin of the cap. **SPORES:** White, elliptical, smooth, amyloid. **HABITAT:** Single or scattered on ground under conifers and aspen in forests, mycorrhizal. **EDIBILITY:** Edible.

The bright yellow scaly cap, yellowish gills, partial veil, and scaly stalk are the important field marks. ***Floccularia albolanaripes*** (*Armillaria albolanaripes*) is very similar, but it has a yellow to yellow-brown cap (becoming brown in the center) with appressed fibrils (in contrast to raised scales on the cap of *F. luteovirens*). It grows singly or scattered under conifers. ***Tricholoma vernaticum*** (p. 169) is identified by a white to brownish gray fibrillose cap, slight ring, distinctive cucumber-like odor, and growth in the mountains under conifers soon after the snow melts. The stalk may be sheathed by veil remnants, but it lacks the conspicuous scales of *Floccularia.*

Laccaria laccata

CAP: 2 to 5 cm in diameter, convex, plane, or depressed in the center, the margin inrolled when young, reddish or orangish

Laccaria laccata

tan, hygrophanous (fading as it loses moisture), fading in age, dry, smooth, odor and taste mild. **GILLS:** Adnate to slightly decurrent, thick and fairly distant, colored like cap but paler, of varying lengths. **STALK:** 2 to 6 cm long, 3 to 7 mm wide, equal, tough and fibrous, colored like cap but darker, often twisted, the base with adhering white mycelium, partial veil absent. **SPORES:** White, subglobose, spiny. **HABITAT:** Scattered, gregarious, or clustered on ground in mycorrhizal association with many kinds of trees in many habitats. **EDIBILITY:** Edible, but the tasteless and tough stalks are usually discarded.

This common mushroom is characterized by the reddish tan color of the cap, which fades to light brown or buff in age, thick gills, white spores, and fibrous stalk. ***Laccaria amethysteo-occidentalis*** is similar in all respects except all parts of the fruiting body are purplish. Even as the color of the cap and stalk fade in age, there usually remains a purple tinge to the gills. ***Laccaria bicolor*** is generally colored like *L. laccata* but has a purplish tint to the gills; if the color of the cap and gills fades to brown, it usually retains a purplish color in the mycelium on and about the base of the stalk. All of these species are common under pines.

Leucopaxillus albissimus

CAP: 5 to 20 cm or more in diameter, convex to plane, the margin inrolled when young, the surface dry, dull, unpolished, off-white becoming cream colored to buff in age, smooth but sometimes cracked in age, firm, flesh white, dry, and brittle, odor mild or unpleasant, taste mild or bitter. **GILLS:** Decurrent, close, white to cream colored to buff in age. **STALK:** 3 to 10 cm long, 2 to 5 cm wide, enlarged toward the base and then often sharply tapered at the point of attachment, firm, solid, colored like the cap, a dense white mycelial mat adhering to the base, partial veil absent. **SPORES:** White, ornamented with amyloid warts. **HABITAT:** Scattered or gregarious, sometimes in arcs and rings, under a variety of trees, including pines, Coast Redwood, and eucalyptus. **EDIBILITY:** Inedible.

The dull, dry, and unpolished cream-colored cap, decurrent gills, and dense mat of white mycelia at the base of the stalk are the important field characteristics. *Leucopaxillus gentianeus* (*Leucopaxillus amarus*) is a close relative that has a reddish brown, unpolished cap, white gills, an unpleasant odor, and a distinctly bitter taste. Both of these robust mushrooms are very firm and decompose very slowly. The decurrent gills might cause confusion with *Clitocybe* species, but those mushrooms are not as firm or durable as species of *Leucopaxillus*.

Leucopaxillus albissimus

Lyophyllum decastes

Lyophyllum decastes
Fried Chicken Mushroom

CAP: 4 to 12 cm in diameter, convex, some shade of brown, from a hint of brown to grayish brown to brown, the margin inrolled at first and often wavy and lobed, surface moist but not viscid, smooth, flesh firm and white, taste and odor mild. **GILLS:** Adnate, notched, or slightly decurrent, white. **STALK:** 5 to 10 cm long, 1 to 3 cm wide, equal, dry, white, smooth, solid, partial veil absent. **SPORES:** White, subglobose, smooth, inamyloid. **HABITAT:** Typically in large clusters in a variety of woods, disturbed areas, parks, gardens, along paths, and so on. **EDIBILITY:** Edible.

The clustered growth habit, grayish brown to brown smooth cap, ringless stalk, and white spores are the principal features. Fruiting bodies are relatively firm, and clusters are often quite heavy. The common name is probably more fanciful than factual. *Lyophyllum semitale,* which also has a brown cap and white stalk, grows singly or in small clusters in coniferous forests. It stains gray to black when bruised. Another grayish to pale brown species that often grows in clusters in disturbed places, especially in wood chips in landscaped area, is *Pluteus petasatus.* Like *L. decastes*, it is often quite robust, but the pink spores and free, crowded gills distinguish it. Because the gills remain white until the fruiting body is mature, *P. petasatus* causes confusion with white-spored mushrooms. It is edible.

Flammulina velutipes
Velvet Foot, Enoki

CAP: 1 to 5 cm in diameter, convex to flat, viscid when moist, brownish orange, odor and taste mild. **GILLS:** Adnexed or notched, pale orange, close. **STALK:** 2 to 8 cm long, 4 to 7 mm wide, equal, pale brownish orange with a dark brown velvety layer of short hairs, partial veil absent. **SPORES:** White, elliptical, smooth, and inamyloid. **HABITAT:** In clusters on many kinds of hardwoods. **EDIBILITY:** Edible.

The clustered growth on wood, viscid or at least sticky orange cap, white spores, and dark pubescence on the lower part of the stalk are a unique combination of features. Commercially, it is grown in the dark in cool temperatures, resulting in a fruiting body with a long, thin white stalk and tiny white cap. It is then sold as Enoki and used in soups and salads. *Flammulina populicola* is similar but grows on aspen. Other species that fruit in clusters include *Connopus acervatus* (*Collybia acervata*), which has reddish to purplish brown hygrophanous caps, reddish brown, cartilaginous stalks, and white spores. It grows in tight bundles on decaying conifers; *Gymnopus villosipes*

Flammulina velutipes

(*Collybia fuscopurpurea*), with dark brown, hygrophanous, deeply striate caps that may become depressed in the center, which grows in clusters or gregariously on the ground under conifers; and **G. erythropus** (*Collybia erythropus*), which has reddish brown to orange-brown caps and stalks. The stalks of all of these species are covered with hairs, especially near the base.

Hygrophoropsis aurantiaca
False Chanterelle

CAP: 2 to 10 cm in diameter, convex to plane, often centrally depressed, orange to brownish orange, often brown in the center, fading to tan, dry, the margin initially incurved, flesh thin and tinged cap color, odor mild. **GILLS:** Decurrent, close, forked several times, bright orange and retaining the orange color even when the cap fades in age. **STALK:** 2 to 8 cm long, 0.5 to 1 cm wide, equal or enlarged at the base, central or eccentric, colored like the cap, partial veil absent. **SPORES:** White, elliptical, smooth, dextrinoid. **HABITAT:** Scattered, gregarious, or clustered on woody debris, especially common under conifers. **EDIBILITY:** Unknown.

The overall orange color, closely spaced, decurrent, forked gills, white spores, and growth on woody debris distinguish this common mushroom. Despite its common name, *H. aurantiaca*

Hygrophoropsis aurantiaca

is not likely to be confused with **Cantharellus californicus** (p. 270), **C. formosus,** or other true chanterelles. The gills of *H. aurantiaca* are a deeper orange, more bladelike, and more closely spaced than the ridges of the true chanterelles. *Cantharellus* species grow on the ground in mycorrhizal association with trees, while *H. aurantiaca* grows as a saprobe on woody debris.

Neolentinus ponderosus

(Other name: *Lentinus ponderosus*)

CAP: 7 to 30 cm in diameter, convex, the center usually slightly depressed in age, white to creamy with buff to brown scales, staining yellow to orange-brown in age or where bruised, surface dry, flesh white, aging or bruising yellow or tan, taste mild, odor fragrant. **GILLS:** Adnate to decurrent, white to cream colored, becoming tan in age with reddish brown stains, edges serrated. **STALK:** 6 to 18 cm long, 2 to 7 cm wide, central or eccentric, tapered toward the base, firm, solid, cream colored aging brown, often with brownish scales, partial veil absent. **SPORES:** White. **HABITAT:** Single or in small clusters at the base of dead conifers or on fallen logs. Fruiting in late spring and summer in mountains. **EDIBILITY:** Edible.

Neolentinus ponderosus

The brown scales on the whitish cap, serrated gills, ringless stalk, and growth on wood are the principal field marks of this large mushroom. It is usually found at high elevations, especially at the base of dead lodgepole pines and ponderosa pines. **Neolentinus lepideus** (*Lentinus lepideus*) is similar but is smaller (cap to 10 cm in diameter) and has a ring on the stalk. Called the "train-wrecker," it was the cause of a brown rot of untreated railroad ties. **Catathelasma imperiale** is one of the largest mushrooms encountered in montane forests. In addition to its large size (up to 40 cm broad), the yellow-brown cap, double-ringed stalk, and very firm texture are important field marks. It has white decurrent gills and a mealy odor.

Omphalotus olivascens
Jack-O-Lantern

CAP: 4 to 18 cm in diameter, convex to plane, often with a depressed center in age, margin inrolled at first becoming upturned and wavy in age, the surface smooth and moist but not viscid, yellow-orange or brownish orange, often with olive tints, flesh yellow-orange, odor and taste mild. **GILLS:** Decurrent, close, orange or yellow-orange with olive tints. **STALK:** 4 to 15 cm long, 1 to 4 cm wide, equal or tapered toward base, central or eccentric, yellow-orange and sometimes tinted with olive, smooth, solid, dry, partial veil absent. **SPORES:** White to pale yel-

Omphalotus olivascens

low, elliptical to subglobose, smooth, inamyloid. **HABITAT:** Typically in clusters at the base of stumps or trunks or from buried wood of hardwood trees and shrubs. **EDIBILITY:** Poisonous, causing gastrointestinal upsets.

The olive orange color, clustered growth on wood, decurrent gills, and white spores are the important field traits of the Jack-O-Lantern, so-called because of its color and ability to glow in the dark. The bioluminescence requires a fresh specimen and a very dark setting to be observed. *Gymnopilus junonius* (p. 245) is similar, but it has a ring on its stalk and rusty orange spores. Chanterelles (*Cantharellus* species) are also orange, but they have shallow ridges rather than gills and grow on the ground, not on wood.

Clitocybe and *Infundibulicybe* Species

The more than 200 species of *Clitocybe* in North America are difficult to characterize as a group. In general, they are dull white, gray, or brown, but a few have other hues. Most are moderate in size, but a few are among the largest mushrooms encountered in western forests. Caps are convex to flattened and very often have depressed centers. Partial and universal veils are lacking. The gills are usually decurrent to some degree, but some are adnexed. Odor and tastes are variable. The spores are mostly white, but some are pale yellow or pink. Some species of the genus are edible, although many are poisonous because of the presence of the toxin muscarine. *Clitocybe* species are decomposers of plant matter. The genus *Infundibulicybe* was created to house species that are genetically only distantly related to *Clitocybe*, despite outward similarities. Some mycologists further place some species into the genus *Lepista*.

a. Fruiting near snowbanks in spring
   b. Cap orange-brown; gills cream colored ....
   ....*Infundibulicybe squamulosa*
   b. Cap gray, buff, or brown
      c. White rhizomorphs present at the base of the stalk ....
      .... *Clitocybe albirhiza*
      c. Rhizomorphs absent but a mycelial mat present ....
      .... *C. glacialis* (see *C. albirhiza*)

a. Found in many habitats
  d. Spores tinged pink
    e. Cap buff to tan .... **C. brunneocephala**
    e. Cap lilac or with lilac tones
      f. Cap less than 8 cm broad; fruiting body tinged lilac or not at all.. **C. tarda**
      f. Cap often greater than 8 cm broad; fruiting body all lilac when young .... **C. nuda**
  d. Spores white
    g. Cap large (often greater than 10 cm broad)
      h. Cap whitish, up to 40 cm broad; odor mealy ....
        .... **C. gigantea**
      h. Cap gray, up to 20 cm broad; odor rancid ....
        .... **C. nebularis**
    g. Cap small or modest (usually less than 10 cm broad)
      i. In grass; cap, gills, and stalk white; odor mild ....
        .... **C. dealbata**
      i. In woods
        j. Cap and stalk blue-green; odor of anise ....
          .... **C. odora** (see *C. fragrans*)
        j. Blue-green colors absent
          k. Odor aniselike; cap buff to tan .... **C. fragrans**
          k. Odor not aniselike; cap pinkish tan ....
            .... ***Infundibulicybe gibba***

Clitocybe albirhiza

CAP: 2 to 8 cm in diameter, convex, plane, or funnel shaped, pale reddish brown becoming buff in age, sometimes zonate, at first covered with a whitish bloom but later smooth, dry, odor mild. **GILLS:** Notched to slightly decurrent, buff. **STALK:** 2 to 6 cm long, 0.5 to 1.5 cm wide, equal, colored like cap and also covered with a white bloom when young, smooth or fibrillose, white rhizomorphs attached to the base, partial veil absent. **SPORES:** White, elliptical, smooth, inamyloid. **HABITAT:** Scattered or clustered under conifers in the mountains in spring and early summer near melting snow. **EDIBILITY:** Unknown.

This is one of the most common snowbank mushrooms—those species that are often associated with melting snow in spring and early summer in mountain forests. The pale brown or buff color and fine bloom (powderlike) on both the cap and

Clitocybe albirhiza

stalk and conspicuous rhizomorphs help identify it. Other gray to brownish snowbank mushrooms include **C. glacialis** (*Lyophyllum montanum*), which has a grayish cap with a whitish bloom, adnexed gills, and a more-or-less equal stalk with a mat of white mycelium at the base. **Melanoleuca angelesiana** (pp. ii–iii) is another look-alike, but it has scant mycelium at the base of the stalk and has warted, not smooth, spores.

Clitocybe brunneocephala
(Other name: *Lepista brunneocephala*)
CAP: 6 to 20 cm in diameter, convex to flat, sometimes with an uplifted and wavy margin in age, surface buff to tan, smooth, lubricious, flesh whitish, odor mild or fragrant, taste mild. **GILLS:** Adnate, notched, or slightly decurrent, close, colored like cap. **STALK:** 4 to 8 cm long, 1 to 4 cm wide, equal or enlarged at the base, colored like cap, dry, solid, smooth, partial veil absent. **SPORES:** Pale pink, elliptical, minutely roughened. **HABITAT:** Scattered or gregarious, often in arcs and rings, in grassy areas, along paths, landscaped areas, vacant lots, and in oak leaf litter in woods. **EDIBILITY:** Edible.

The lubricious (slippery or slick but not viscid) texture of the cap surface, pinkish spores, rather short stalk in relation to the

Clitocybe brunneocephala

size of the cap, and lack of a ring on the stalk characterize *C. brunneocephala*. It is very similar to **C. nuda** (p. 151), the Blewit, but it lacks any hint of purple or lilac. **Clitocybe tarda** (p. 152) is smaller (cap 2 to 8 cm broad) and has a brownish cap with lilac tones. It grows in grassy areas and on piles of rotting vegetation, often in arcs and fairy rings.

Clitocybe dealbata
Sweat-producing Clitocybe

CAP: 2 to 4 cm in diameter, convex becoming plane to shallowly depressed in age, white, whitish gray, or buff, smooth, flesh white to cream colored, odor mild. **GILLS:** Adnate becoming slightly decurrent, white to tan, close. **STALK:** 2 to 4 cm long, 3 to 8 mm wide, equal or tapered below, colored like cap, partial veil absent. **SPORES:** White, elliptical, smooth, inamyloid. **HABITAT:** Scattered or in rings and arcs in grassy areas. **EDIBILITY:** Poisonous, causing sweating, salivation, tear flow, diarrhea, abdominal pains, and other unpleasant symptoms, due to the toxin muscarine (in higher concentrations than in *Amanita muscaria*).

The whitish cap, decurrent gills, white spores, ringless stalk, and growth in grass are the distinguishing features of this small mushroom. Its notoriety is due to the dangerous levels of muscarine and potentially serious effects on children who might

Clitocybe dealbata

come in contact with it in lawns. It is sometimes fairly common alongside other grass inhabitants such as **Marasmius oreades** (p. 188) and **Agaricus campestris** (p. 224). *Marasmius oreades* has a tan cap, widely spaced gills that are adnexed or adnate as opposed to decurrent, and the ability to rehydrate after drying, which is not a characteristic of *C. dealbata. Agaricus campestris* has brown spores and free gills.

Clitocybe fragrans
(Other name: *Clitocybe deceptiva*) Anise Mushroom

CAP: 1 to 5 cm in diameter, convex becoming plane or with a depressed center, margin inrolled when young, cap surface smooth, hygrophanous (fading as it loses moisture), buff to tan, odor of anise, taste mild. **GILLS:** Adnate to slightly decurrent, white or buff. **STALK:** 2 to 6 cm long, 2 to 5 mm wide, more-or-less equal or enlarged at the base, whitish or buff, partial veil absent. **SPORES:** Pale pinkish buff. **HABITAT:** Scattered, gregarious, or in rings on ground under conifers. **EDIBILITY:** Edible but has poisonous look-alikes.

The buff to tan cap, modest size, attached gills, and especially the anise odor are the principal field marks. The odor sets it apart from many other modestly sized, light-colored *Clitocybe* species, which are numerous and diverse in western forests. **Clitocybe odora** (p. 424) also has a distinctive aniselike odor, but

Clitocybe fragrans

all parts of the fruiting body are an attractive bluish green. It grows on the ground in hardwood/coniferous forests but never in large numbers. **Clitocybe dealbata** (p. 147), a poisonous species, is similar to *C. fragrans*, but it lacks the anise odor and grows in grassy areas such as lawns and pastures.

Clitocybe gigantea
(Other name: *Leucopaxillus giganteus*)
CAP: 10 to 40 cm in diameter, convex to plane, becoming depressed in the center, white aging buff or tan, the margin inrolled at first, the surface smooth, dry or moist but not viscid, fragile in age, odor and taste mealy or unpleasant. **GILLS:** Decurrent, crowded, whitish but soon cream to buff, sometimes forked. **STALK:** 3 to 10 cm long, 2 to 5 cm wide, more-or-less equal, white, dry, smooth, sometimes with white mycelia adhering to the base. **SPORES:** White, elliptical, smooth, slightly amyloid. **HABITAT:** Scattered or gregarious or often in arcs and rings, sometimes in large numbers, in mountain forests of the West. **EDIBILITY:** Unknown.

This is a very large whitish mushroom that is often encountered growing in large fairy rings in the Rocky Mountains and other mountainous regions of the West. The caps are often dinner-plate size and bright white in the gloom of the forest floor. It might be confused with **Leucopaxillus albissimus** (p. 138), but that species has a dull, mattelike texture of the surface of the cap,

Clitocybe gigantea

a dense mat of mycelium that adheres to the base of the stalk, and a tough, not fragile, fruiting body. ***Clitocybe maxima*** (a "borrowed" European name of doubtful accuracy when applied to mushrooms in the western United States) is also very large. It has a deeply funnel-shaped, pinkish tan cap and a thick whitish stalk. It is also found in mountains of the West.

Clitocybe nebularis

CAP: Up to 20 cm or more in diameter, at first rounded, becoming convex to plane or with a slightly depressed center, gray with a whitish bloom when young becoming grayish brown in age, the margin inrolled at first, wavy in age, and paler than the center of the cap, odor unpleasant, rancid. **GILLS:** Adnate to decurrent, white to buff, close. **STALK:** 5 to 10 cm long, 2 to 4 cm wide, enlarged toward the base, white to creamy or buff, smooth or minutely fibrillose, with white downy hairs covering the base. **SPORES:** Pale yellow, elliptical, smooth, inamyloid. **HABITAT:** Scattered, gregarious, or in arcs and rings in hardwood/coniferous forests, especially common in central and northern California coastal forests. **EDIBILITY:** Edible according to some sources, but the obnoxious odor is a deterrent.

This robust mushroom has a distinctly unpleasant odor, grayish brown cap, decurrent gills, and white spores. When

Clitocybe nebularis

young, the rounded cap and adnate gills mask its identity since *Clitocybe* species typically possess decurrent gills. Mature specimens might be confused with **Leucopaxillus albissimus** (p. 138), which also may have an unpleasant odor, but that species has a firmer texture and white spores, or **Tricholoma saponaceum** (p. 166), which has a gray or greenish brown cap and an unpleasant soaplike odor and lacks decurrent gills. The base of its stalk is pink or orange, a feature not found in *C. nebularis.*

Clitocybe nuda

(Other name: *Lepista nuda*) Blewit

CAP: 4 to 15 cm in diameter, convex to flat, bluish, purple, or lilac, losing the purple and fading to tan in age, the margin inrolled at least when young, often wavy in age, surface smooth, lubricious, tacky when moist, flesh lilac, odor fragrant, taste mild. **GILLS:** Adnexed, notched, or slightly decurrent, lilac, fading to tan. **STALK:** 2 to 6 cm long, 1 to 2.5 cm wide, equal or enlarged at the base, colored like cap, dry, finely fibrillose, the base with adhering purplish mycelium, partial veil absent. **SPORES:** Pale pink, elliptical, roughened. **HABITAT:** Single, scattered, or gregarious on ground, saprobic on decomposing organic matter in many habitats, such as woods, landscaped areas, and gardens. **EDIBILITY:** Edible and choice.

Clitocybe nuda

The purple color of all parts of the fruiting body, pink spores, absence of a ring, and smooth, lubricious cap separate the Blewit from other purplish mushrooms such as **Laccaria amethysteo-occidentalis** (with white spores and a very fibrous stalk) and several *Cortinarius* species (with rusty brown spores and a zone of hairs on the stalk). **Clitocybe brunneocephala** (p. 146) is very similar, but all parts of the fruiting body are tan or buff. **Clitocybe tarda** (below) is smaller (cap 2 to 8 cm broad) and has a brownish cap with lilac tones. **Lepista subconnexa** (*Clitocybe subconnexa*) has a white to buff cap, pinkish tinged spores, clustered habit, and a pleasant odor.

Clitocybe tarda
(Other name: *Lepista tarda*)

CAP: 2 to 8 cm in diameter, convex becoming flat to funnel shaped, the margin inrolled at first, often becoming upturned and wavy in age, surface smooth, lubricious, grayish brown with or without a tinge of lilac, fading to tan as it ages and dries, odor mild or fragrant. **GILLS:** Adnate to decurrent, close, colored like cap. **STALK:** 2 to 6 cm long, 3 to 8 mm wide, equal, colored like cap, fibrillose, partial veil absent. **SPORES:** Pale pink, elliptical,

Clitocybe tarda

minutely roughened. **HABITAT:** Gregarious or clustered, often in arcs and rings in lawns and other grassy areas and on piles of rotting vegetation. **EDIBILITY:** Unknown.

This modestly sized mushroom is recognized by the grayish brown cap that may have a tinge of lilac, pink spores, and occurrence in grassy areas, where it sometimes forms large fairy rings. It is reminiscent of **C. nuda,** the Blewit (p. 151), but the lilac tinge of *C. tarda*, if present at all, is subtle. The Blewit, on the other hand, is distinctly lilac or purple and usually retains some lilac even in old age. **Clitocybe brunneocephala** (p. 146), which sometimes grows in fairy rings in grassy areas, is tan and significantly larger than *C. tarda.*

Infundibulicybe gibba
(Other name: *Clitocybe gibba*)
CAP: 3 to 9 cm in diameter, strongly funnel shaped, smooth, dry, tan or pinkish tan, margin straight or wavy, odor mild or sweet. **GILLS:** Decurrent, crowded, white or creamy. **STALK:** 2 to 8 cm long, up to 1 cm wide, more-or-less equal, white or colored like the cap but paler, dry, the base covered with white downy mycelium. **SPORES:** White, teardrop shaped, smooth, inamyloid. **HABI-**

Infundibulicybe gibba

TAT: Single or scattered on ground, usually under hardwoods, especially oaks. **EDIBILITY:** Edible, but similar appearing species are poisonous.

The funnel-shaped pinkish tan cap that usually contrasts sharply with the white decurrent gills helps to distinguish this widespread species. Two species are similar in some respects. *Infundibulicybe squamulosa* (below) has an orange-brown cap and stalk and cream-colored gills and grows under conifers soon after the snow melts in the high mountains. **Pseudoclito-cybe cyathiformis** (*Clitocybe cyathiformis*) has a brownish cap that lacks the pinkish tint of *I. gibba* but otherwise is quite similar, and microscopic examination is necessary to separate the two species. Unlike *I. gibba*, *P. cyathiformis* has spores that turn blue (amyloid) in Melzer's reagent. In addition to these two species, western forests have many other pale, funnel-shaped *Clitocybe* species. Some grow on the ground, whereas others grow directly on rotted wood.

Infundibulicybe squamulosa
(Other name: *Clitocybe squamulosa*)
CAP: 2 to 9 cm in diameter, funnel shaped, orange-brown but fading in age, at first with minute scales on the surface but these

quickly disappear, resulting in a smooth appearance, flesh white; odor and taste mild or faintly mealy. **GILLS:** Decurrent, creamy, often forked. **STALK:** 3 to 8 cm long, 0.5 to 1 cm wide, colored like cap, equal, dry, the base covered with white hairs. **SPORES:** White, elliptical, smooth, inamyloid. **HABITAT:** Single or scattered under conifers in the mountains, often near melting snow. **EDIBILITY:** Unknown.

This is one of the snowbank mushrooms common in the higher mountains in the western United States. The genus *Infundibulicybe* was proposed because, based on DNA evidence, this species and others are only distantly related to *Clitocybe* species. Nevertheless, the conspicuously decurrent gills and white spores immediately suggest a *Clitocybe*. It is an attractive mushroom with cream-colored gills sandwiched between an orange-brown cap and identically colored stalk. Other moderately sized snowbank mushrooms include **Clitocybe albirhiza** (p. 145) and **C. glacialis,** but those species have duller gray to brown caps. **Clitocybe salmonilamella** has a centrally depressed, grayish orange cap, pinkish gills, and a sweet odor. The cap fades to whitish in age. It grows on rotted wood or on soil rich in decayed wood material.

Infundibulicybe squamulosa

Clitopilus prunulus
Sweetbread Mushroom

CAP: 2 to 10 cm in diameter, convex becoming plane, the center often depressed, dry or slightly viscid when moist, white or tinged gray, the margin inrolled at first, later often wavy, flesh white, odor distinctly mealy, taste mild. **GILLS:** Decurrent, close, white or gray becoming faintly pink as the spores mature. **STALK:** 2 to 8 cm long, 7 to 15 mm wide, central or eccentric, equal or tapering toward the base, solid, dry, white or tinged gray, the base covered with white mycelium, partial veil absent. **SPORES:** Pink, elliptical, longitudinally ridged, angular in end view. **HABITAT:** Single or scattered on ground in hardwood/coniferous forests, typically fruiting early in the mushroom season. **EDIBILITY:** Edible but resembles some toxic species.

The mealy or sweetbread odor together with the white or pale gray color of the fruiting body and pink spores are the defining field marks. Its cap color is reminiscent of ***Clitocybe dealbata*** (p. 147), a poisonous mushroom found in lawns, but that mushroom has white spores and lacks the odor of *C. prunulus*. Numerous white to gray woodland species of *Clitocybe* are not covered in this field guide. However, the pinkish spores and especially the strong odor of the Sweetbread Mushroom are distinctive.

Clitopilus prunulus

Melanoleuca melaleuca

Melanoleuca melaleuca

CAP: 2 to 8 cm in diameter, convex but soon plane with or without a low umbo, gray to dark brown becoming tan in age, odor and taste mild. **GILLS:** Notched, close, white. **STALK:** 2 to 8 cm long, up to 1 cm wide, equal or enlarged at the base, straight, stiff, white or light brown with brown fibrils, partial veil absent. **SPORES:** White, elliptical, warted, amyloid. **HABITAT:** Single or scattered on ground in a variety of habitats—in landscaped areas, lawns, and pastures, in wood chips, along paths, and in forests. **EDIBILITY:** Edible.

This is a highly variable species that makes a positive identification difficult; moreover, the name itself may not be appropriate for all western collections. However, a species fitting the above description is commonly encountered in many habitats, including urban areas. In general, the cap is usually brownish but may be gray or pale tan when dry or growing in direct sunlight; the stalk is stiff and straight, and the gills and spores are white. ***Melanoleuca angelesiana*** (pp. ii–iii) is common in montane forests soon after the snow melts. It is similar to the description above, but it lacks cystidia on the gills, which are present in *M. melaleuca*. The cap is grayish brown but fades to a nondescript grayish tan.

Tricholomopsis rutilans
Plums and Custard

CAP: 3 to 12 cm in diameter, convex to nearly flat, dry, conspicuously bicolored from dense purplish red or vinaceous fibrils on a yellow background, the fibrils denser toward the center of the cap, flesh yellow, odor and taste mild or slightly radishlike. **GILLS:** Notched or adnate, close, yellow. **STALK:** 5 to 10 cm long, 1 to 2.5 cm wide, equal, dry, yellow with purplish red fibrils (but less densely covered than the cap), partial veil absent. **SPORES:** White, elliptical, smooth, inamyloid. **HABITAT:** Single, scattered, or in small clusters on decaying coniferous wood or on lignin-rich soil. **EDIBILITY:** Edible.

The combination of purple-red scales on a yellow ground color plus the yellow gills result in a very attractive mushroom that is not likely to be confused with any other species. It doesn't

Tricholomopsis rutilans

seem to fruit in large numbers. **Megacollybia fallax** (*Tricholomopsis fallax*) is similar in stature but has a grayish brown to brown cap up to 10 cm broad. The surface of the cap often develops a streaked appearance from radiating flattened fibrils. The gills are white and notched, and the stalk is whitish with brown fibrils. White rhizomorphs are attached to the base of the stalk. It is found in western mountains on or near rotting wood.

Tricholoma Species

This is a large genus in the western United States. As a group, they share white spores, notched gills, and a mycorrhizal livelihood. Otherwise, species vary in size from small to large and in cap texture from smooth to scaly. A ring on the stalk may or may not be present. Odors are also variable; some are quite distinctive and an important characteristic for making species determinations in the field.

a. Ring present
  b. Odor spicy; cap and stalk white, often with brownish stains .... **Tricholoma magnivelare**
  b. Odor cucumber-like; cap white to brownish gray; fruiting soon after snowmelt .... **T. vernaticum**

a. Ring absent
  c. Cap yellow; stalk pale yellow................................ **T. equestre**
  c. Cap not primarily yellow but may be cream colored
    d. Cap typically less than 4 cm broad, gray ..... **T. myomyces**
    d. Cap typically more than 4 cm broad, variously colored
      e. Base of stalk pinkish; odor soapy .... **T. saponaceum**
      e. Base of stalk not pinkish; odor mild or mealy
        f. Cap viscid when moist
          g. Cap creamy to cinnamon brown ....**T. dryophilum** (see *T. imbricatum*)
          g. Cap violet-gray........................**T. griseoviolaceum**
        f. Cap dry
          h. Cap white with tiny gray scales........**T. pardinum**
          h. Cap brown, fibrillose or scaly; stalk solid............ .... **T. imbricatum**
          h. Cap reddish brown, fibrillose to scaly; stalk hollow ....**T. vaccinum**

Tricholoma dryophilum

Tricholoma equestre

(Other name: *Tricholoma flavovirens*) Man on Horseback

CAP: 5 to 15 cm in diameter, convex to nearly flat, sometimes with an upturned margin in age, entirely yellow or yellow with a brownish center, viscid in moist weather, smooth, flesh white,

Tricholoma equestre

odor mealy. **GILLS:** Notched, close, yellow. **STALK:** 4 to 10 cm long, 1 to 4 cm wide, equal or clavate, dry, smooth, pale yellow discoloring yellow-brown at the base, partial veil absent. **SPORES:** White, elliptical, smooth. **HABITAT:** Scattered to gregarious on ground in mycorrhizal association with conifers, especially pines, and some hardwoods, such as aspen. **EDIBILITY:** Previously considered edible, but recent reports of poisonings have occurred.

The yellow color of the cap, notched yellow gills, and ringless stalk are the principal field marks. In age, the cap may develop brownish tones, especially on the center. A few other yellow mushrooms of forests include *T. sejunctum,* which has radiating, appressed blackish fibrils on the center of the cap and white gills; *T. sulphureum,* with a very strong repulsive odor (coal-tar-like) and a dry yellow cap and stalk; and *T. intermedium,* which has a viscid greenish yellow cap, white gills, and a white stalk. *Floccularia luteovirens* (p. 135) has a yellow cap with conspicuous scales and a cottony ring on a scaly stalk; *F. albolanaripes* (p. 136) is similar but has a yellow to yellow-brown cap with appressed fibrils.

Tricholoma griseoviolaceum

CAP: 4 to 10 cm in diameter, umbonate at first, becoming convex to nearly flat, the margin often wavy and sometimes splitting in age, surface viscid when moist, radially streaked, violet-gray with a darker center, becoming brownish gray in age with a paler margin, flesh whitish to gray, odor and taste cucumber-like. **GILLS:** Notched, white, discoloring pinkish brown in age. **STALK:** 6 to 12 cm long, 1 to 2 cm wide, equal to tapered at the base, dry, sparingly fibrillose, white, partial veil absent. **SPORES:** White, elliptical, smooth, inamyloid. **HABITAT:** Single, scattered, or gregarious on ground, associated with oak and tanoak. **EDIBILITY:** Unknown.

Field marks of *T. griseoviolaceum* include the viscid, violet-gray cap that is radially streaked with dark gray or black, white notched gills, white stalk, and an association with hardwoods. It is most easily confused with *T. portentosum,* which is characterized by a viscid, dark gray to grayish brown cap with appressed, radiating gray fibrils, yellow tones that often develop on the cap and stalk, and an association with conifers. Yellow colors are

Tricholoma griseoviolaceum

absent in *T. griseoviolaceum*. Other violet or gray *Tricholoma* species associated with hardwoods include **T. virgatum** (p. 424), which has a dry, streaked, gray, acutely umbonate cap, and **T. atroviolaceum,** which has a dry, fibrillose, dark violet-gray cap and grayish gills.

Tricholoma imbricatum

CAP: 6 to 18 cm in diameter, convex to nearly flat, brown, dry, smooth when young but soon covered with flattened fibrils or fibrillose scales, flesh firm, white, odor and taste mild. **GILLS:** Notched or adnexed, close, white, developing brown edges in age. **STALK:** 4 to 12 cm long, 2 to 3 cm wide, equal or clavate, white to buff, becoming brown in age, dry, solid, partial veil absent. **SPORES:** White, elliptical, smooth. **HABITAT:** Single to gregarious on ground under conifers, especially common in coastal forests. **EDIBILITY:** Unknown.

This robust *Tricholoma* is recognized by the brown fibrillose cap, notched gills, white spores, and absence of a ring on the stalk. The gills may be tinged brown in age. Similar *Tricholoma* species include **T. vaccinum** (p. 167), which has a reddish brownish and scaly cap, hollow whitish stalk, and whitish gills that often develop brown spots. It is common under conifers in

Tricholoma imbricatum

the high mountains but has a wider distribution. **Tricholoma dryophilum** (p. 160) is white to cream colored when young but develops cinnamon brown streaks in age and may become predominately brown. It fruits on the ground under oaks. It lacks a partial veil. **Tricholoma manzanitae** has a pale orange-brown cap and fruits under manzanita; **T. fracticum** has a brown viscid cap, two-colored stalk (whitish above the slight ring, brown below), and bitter taste. It fruits under conifers.

Tricholoma magnivelare

(Other name: *Armillaria ponderosa*) Matsutake

CAP: 5 to 20 cm in diameter, convex to plane, the margin inrolled at first, white but soon spotted with reddish yellow or reddish brown stains, dry to slightly viscid when moist, odor fragrant, spicy, or pinelike. **GILLS:** Adnate to notched, white, developing brown stains in age, crowded. **STALK:** 4 to 15 cm long, 2 to 6 cm wide, firm, solid, equal, white with brown stains, the partial veil leaving a large membranous cottony ring. **SPORES:** White, elliptical, smooth. **HABITAT:** Single, scattered, or in arcs and rings on ground in mycorrhizal association with pines, tanoaks, and madrone. **EDIBILITY:** Edible and choice, but its texture is rather tough.

Tricholoma magnivelare

The Matsutake is one of the most sought-after mushrooms in the Pacific Northwest, where it is harvested commercially and often exported to Asia. The white color of the cap and stalk, which are usually spotted with reddish brown stains, presence of a ring, and especially the spicy odor are the principal field marks. It has a firm texture and is often bug free. Like many mushrooms, it often fruits just below the surface of the duff, so keen-eyed investigation is needed. ***Tricholoma focale*** (*T. zelleri*) has a yellow-brown or orange-brown cap, whitish gills, orange-brown stains on the gills and stalk, a membranous ring, and a mealy odor; ***T. caligatum*** (*Armillaria caligata*) has a white cap overlaid with brown fibrils and a spicy odor but a very bitter taste; and ***T. vernaticum*** (p. 169), a robust snowbank mushroom, is readily identified by a cucumber-like odor and fibrillose white to grayish cap.

Tricholoma myomyces
(Other name: *Tricholoma terreum*)

CAP: 2 to 5 cm in diameter, convex to nearly flat, dry, gray from a dense layer of gray or black feltlike fibrils, background color pale gray, the margin usually paler than the rest of the cap, odor

Tricholoma myomyces

mild, taste mealy. **GILLS:** Notched, close, white or tinged gray. **STALK:** 2 to 5 cm long, 0.5 to 1 cm wide, more-or-less equal, white or tinged gray, mostly smooth, partial veil fibrillose, leaving remnants on the stalk or margin of the cap but soon disappearing. **SPORES:** White, elliptical, smooth, inamyloid. **HABITAT:** Scattered or gregarious on ground in coastal forests under conifers. **EDIBILITY:** Unknown.

The small size of the dry gray cap covered with grayish fibrils, notched whitish gills, white spores, and growth on the ground help identify this mushroom. ***Tricholoma moseri*** is another small *Tricholoma* with a densely fibrillose or minutely scaly cap. The surface of the cap is grayish to blackish brown, the stalk is white to gray and up to 5 cm long, and the gills are gray. Unlike *T. myomyces*, it fruits in montane forests under conifers soon after the snow melts in spring.

Tricholoma pardinum

Tiger Tricholoma

CAP: 5 to 15 cm in diameter, convex to nearly flat, dry, white to pale gray with scattered dark grayish or brownish gray tiny scales, the margin paler relative to the center, flesh white to light gray, odor and taste mealy. **GILLS:** Notched, whitish to cream col-

Tricholoma pardinum

ored. **STALK:** 4 to 12 cm long, 1 to 3 cm wide, equal or enlarged at the base, white or cream colored, dry, silky-fibrillose, the fibrils of the lower stalk staining brownish, partial veil absent. **SPORES:** White, elliptical, smooth, inamyloid. **HABITAT:** Scattered or gregarious on ground in hardwood/coniferous forests, especially under conifers, widespread. **EDIBILITY:** Poisonous, causing gastrointestinal upsets.

This is one of the robust species of *Tricholoma*. Its dry, whitish cap with scattered or regularly spaced dark tiny scales separates it from other *Tricholoma* species with nonviscid caps: **T. imbricatum** (p. 163), which has a brown cap matted with fibrils; **T. virgatum** (p. 424), with a gray-streaked cap that is conical when young and typically acutely umbonate in age, plus an acrid taste; and **T. atroviolaceum,** which has a violet-gray cap that is almost black at the center. The flesh of the cap of *T. atroviolaceum* bruises reddish gray, a feature apparently unique among *Tricholoma* species.

Tricholoma saponaceum
Soapy Tricholoma
CAP: 4 to 15 cm in diameter, convex to nearly flat, sometimes with a broad umbo, the margin inrolled at first, often uplifted and wavy in age, surface dry or moist but not viscid, smooth but

Tricholoma saponaceum

sometimes developing cracks in dry weather, color highly variable: gray, olive, grayish brown, grayish olive, or yellowish brown, the margin often paler, flesh whitish, odor and taste soapy. **GILLS:** Notched, white or cream colored. **STALK:** 4 to 12 cm long, 1 to 2.5 cm wide, equal but tapered at the base, smooth, white tinged gray or brown, pale pink or orange at the base, flesh whitish but the base interior tinted pinkish or orange, partial veil absent. **SPORES:** White, elliptical, smooth, inamyloid. **HABITAT:** Single, scattered, or in small clusters on ground in hardwood/coniferous forests. **EDIBILITY:** Unknown.

The color of the cap of this very common *Tricholoma* is highly variable, ranging from gray to olive to brown. The smooth texture of the surface of the cap, notched gills, white spores, and especially the pinkish coloration of the base of the stalk and the soapy odor are the principal field marks. For other species of *Tricholoma* with grayish caps, see *T. griseoviolaceum* (p. 161).

Tricholoma vaccinum

CAP: 3 to 8 cm in diameter, convex to nearly flat in age, often with a rounded umbo, the margin inrolled at first and edged with fibrils from the partial veil, the surface reddish brown to vinaceous with a darker center, fading in age, the surface fibrillose becoming scaly in age, dry, sometimes cracking in dry weather, flesh white, slowly bruising pink, odor and taste mild or mealy. **GILLS:** Notched, close, white, developing brownish spots in age. **STALK:** 3 to 8 cm long, 1 to 2 cm wide, equal to clavate, whitish, buff, or yellow-brown with brown fibrils and scales, dry, hollow in age, partial veil absent. **SPORES:** White, elliptical, smooth, inamyloid. **HABITAT:** Single to gregarious on ground under conifers in mountains. **EDIBILITY:** Unknown.

Tricholoma vaccinum is recognized by a fibrillose to scaly, reddish brown cap, white gills, and hollow stalk. Other brownish *Tricholoma* species include *T. imbricatum* (p. 163), which has a brown fibrillose cap and solid stalk; *T. dryophilum* (p. 160), cream colored when young but developing cinnamon brown streaks, with a preference for oaks; *T. muricatum* (*T. pessundatum*), which has a reddish brown cap finely streaked from radiating fibrils, cream-colored gills that develop brownish stains, a buff-colored stalk that discolors brown in age, and a preference for pines; and *T. fractium*, which has a brown viscid cap, two-colored stalk (whitish above the slight ring, brown below), and bitter taste and fruits under conifers.

Tricholoma vaccinum

Tricholoma vernaticum

(Other name: *Armillaria olida*)

CAP: 5 to 15 cm in diameter, convex to broadly umbonate to nearly flat, often with an uplifted and wavy margin in age, whitish to gray or grayish brown, surface viscid when moist, smooth when young becoming fibrillose, flesh white, odor like cucumbers or raw potatoes. **GILLS:** Notched, close, white or tinged gray, of different lengths. **STALK:** 5 to 14 cm long, 2 to 4 cm wide, equal or swollen at the base, dry, fibrillose, white or tinged ochre to reddish brown, firm and solid, partial veil leaving a fibrillose thin ring that may disappear. **SPORES:** White, elliptical, smooth, inamyloid. **HABITAT:** Single, gregarious, or in small clusters in duff under conifers, fruiting in spring soon after snow melt. **EDIBILITY:** Unknown; the odor may be a deterrent.

This common snowbank mushroom is characterized by the white to brownish gray cap, white spores, slight ring, and especially the cucumber or raw-potato odor. It is often encountered in spring in the Sierra Nevada pushing up conifer duff in mushroom shrumps. ***Hygrophorus subalpinus*** is similar and also common in the mountains under conifers near melting snow. It has a white viscid cap, white stalk, an ephemeral ring, and a mild odor. It is distinguished from *T. vernaticum* by its waxy, adnate to decurrent gills and mild odor.

Tricholoma vernaticum

Xeromphalina campanella

CAP: 0.5 to 2 cm in diameter, convex becoming nearly flat with a depressed center (cover photograph), reddish to orange-brown, smooth, the margin incurved at first, striate, odor and taste mild. **GILLS:** Decurrent, fairly well spaced, often with many cross-veins, yellow to dull orange. **STALK:** 1 to 5 cm long, 1 to 3 mm wide, equal or enlarged at the base, tough and somewhat wiry, smooth, reddish brown with a paler yellowish apex, the base covered with brown hairs, partial veil absent. **SPORES:** White, elliptical, smooth, amyloid. **HABITAT:** Scattered or in dense clusters on decaying coniferous wood. **EDIBILITY:** Unknown.

This small mushroom, which is usually found fruiting in large numbers, is limited to decaying conifers. The reddish brown cap, yellowish gills, white spores, and tough thin stalk are the principal field marks. Similar species include **X. cauticinalis**, which grows on conifer needles and debris, and **X. fulvipes**, which also grows on conifer needle duff but has adnate gills (rather than decurrent gills) and a bitter taste. Microscopic details separate these and other species of *Xeromphalina*. Other common mushrooms that grow on wood or on conifer duff include species of *Mycena*, but they general have long, delicate stalks.

Xeromphalina campanella

Lichenomphalia umbellifera

(Other name: *Omphalina ericetorum*)

CAP: Up to 3 cm in diameter, at first plane with an incurved margin, becoming funnel shaped, yellow-brown, fading to yellow in age, smooth, margin striate and often wavy. **GILLS:** Decurrent, distant, pale yellow fading to whitish. **STALK:** 1 to 3 cm long, 1 to 3 mm wide, colored like cap above and brownish below, equal or enlarged at the base, often curved, smooth, partial veil absent. **SPORES:** White to yellowish, elliptical, smooth, inamyloid. **HABITAT:** Scattered or gregarious on rotting conifer logs and stumps and on damp soil; associated with lichen. **EDIBILITY:** Unknown.

This mushroom is characterized by its small size, funnel-shaped yellowish cap, decurrent gills, and association with lichens. Similar mushrooms include **Contumyces rosellus**, which is pale salmon pink and grows in moss and grassy areas and along trails; **Chrysomphalina aurantiaca** (*Omphalina luteicolor*), which is completely orange and grows on rotting conifer wood in groups or clusters; **Chrysomphalina chrysophylla**, with a fibrillose, yellow-brown cap and stalk that also occurs on decayed conifer wood; and **Chromosera cyanophylla**, which is lilac at first but ages yellow (the gills retaining the lilac color the longest) and grows on or under bark of rotting logs. The unrelated **Rickenella fibula** has an orange cap, relatively long stalk,

Lichenomphalia umbellifera

and cream-colored gills (microscopic examination of cylindrical cystidia helps distinguish it) and grows in moss; interestingly, it is related to crust fungi.

Strobilurus trullisatus

CAP: 0.5 to 1.5 cm in diameter, convex to plane, smooth to wrinkled, the margin striate, whitish with pink or tan tones. **GILLS:** Adnate to adnexed, white to pale pink. **STALK:** 2 to 5 cm long, 1 to 2 mm wide, equal, dry, white above, yellow-brown below and densely and finely hairy, partial veil absent. **SPORES:** White, elliptical, smooth, inamyloid. **HABITAT:** Scattered on partially buried or decayed Douglas-fir cones in moist humid areas. **EDIBILITY:** Unknown but too small to be of interest.

In appearance, this small mushroom resembles a *Mycena*, but its habitat is restricted to fallen cones, especially those of Douglas-fir. It is the most common cone mushroom in the Pacific Northwest. **Strobilurus occidentalis** has a grayish brown cap without a pink cast and prefers spruce cones. **Baeospora myosura** is another common cone dweller that is frequently encountered on spruce cones. It has a smooth brown or pinkish brown cap, crowded gills, brown, hairy stalk base, and weakly amyloid spores. These are not the only mushrooms encountered on conifer cones, but these three species are unique because they only grow on cones.

Strobilurus trullisatus

Caulorhiza umbonata

Caulorhiza umbonata
Redwood Rooter

CAP: 5 to 15 cm in diameter, conical becoming sharply umbonate, pale reddish to yellowish brown, hygrophanous (fading as it loses moisture), surface smooth, margin often wavy in age, odor and taste mild. **GILLS:** Adnexed or notched, close, cream colored or buff. **STALK:** 6 to 50 cm or more long (most below ground), 0.5 to 2 cm wide, tapered downward and extending deep into the substrate, sometimes twisted or flattened, striate or furrowed, smooth or fibrillose, colored like cap but paler, cartilaginous, partial veil absent. **SPORES:** White, elliptical, smooth, amyloid. **HABITAT:** Single or scattered on ground under Coast Redwoods. **EDIBILITY:** Unknown.

The association with redwoods, umbonate cap, white spores, and the very long stalk that is deeply rooted distinguish this mushroom. The taproot of the stalk will usually break off when the mushroom is picked. Coast redwoods are not generally a good habitat for medium to large mushrooms, but *C. umbonata* is an exception. ***Caulorhiza hygrophoroides*** is similar but has a smaller cap (less than 5 cm broad) and pinkish tinged gills. It grows under hardwoods in the Rocky Mountains. *Phaeocollybia* species (p. 250) also have a "taproot," but they have rusty brown spores and viscid caps. They grow under various conifers.

Mycena Species

These mushrooms are generally small, slender, and fragile. The caps are usually conical or bell shaped and striate, the stalks are long and thin, and the spores are white. Some grow in great numbers on the forest floor in leaf litter, whereas others grow on wood. They are too small to be of value for the table.

a. Growing on the ground (including fallen leaves, needles, etc.)
  b. Odor bleachlike; cap gray-brown ..... ***Mycena leptocephala***
  b. Odor not bleachlike
    c. Cap and stalk with lilac or purple hues
      d. Fruiting on buried pine cones and other woody debris
      .... ***M. purpureofusca***
      d. Fruiting on the ground......................................***M. pura***
    c. Cap and stalk without lilac tones
      e. Cap typically less than 1 cm broad
        f. Cap yellow .... ***M. oregonensis***
        f. Cap red....................***M. acicula*** (see *M. oregonensis*)
      e. Cap typically more than 1 cm broad
        g. Cap and stalk yellow...........................***M. epipterygia***
        g. Cap brown; stalk red and yellow .... ***M. nivicola***
        g. Cap reddish or orange; stalk exuding red juice
        when cut......***M. californiensis*** (see *M. haematopus*)

Mycena californiensis

a. Growing on wood (bark, logs, fallen branches, etc.)

   h. Base of stalk exuding red juice when cut; cap reddish..........
   ....**M. haematopus**

   h. Base of stalk not exuding red juice when cut; cap brownish
   .... **M. maculata**

Mycena epipterygia

CAP: 1 to 2 cm in diameter, convex to bell shaped, yellow becoming pale yellow or white near the margin in age, viscid when moist, striate, the margin often toothed or scalloped, flesh thin and yellow, odor and taste mild. **GILLS:** Adnate to slightly decurrent, white to pale yellow. **STALK:** 4 to 9 cm long, 1 to 2 mm wide, equal, fragile, smooth, viscid when moist, yellow becoming whitish in age, partial veil absent. **SPORES:** White, elliptical, smooth, amyloid. **HABITAT:** Scattered to gregarious on ground under conifers. **EDIBILITY:** Unknown.

The yellow viscid cap and stalk separate this *Mycena* from its relatives. Other common *Mycena* species or *Mycena*-like species growing on the ground in forests include **M. aurantiidisca** (pp. xii–1), which has an orange to yellow-orange cap that is often

Mycena epipterygia

paler near the margin and a stalk that is white or white with a yellow base; **M. clavicularis**, with a gray cap, pale gray gills, and a viscid dark gray stalk; **Roridomyces roridus** (*M. rorida*), with a brownish dry cap, white gills, and a very slimy whitish stalk; and **M. galopus**, which has a dark grayish brown, translucent-striate cap that fades to white and a grayish brown stalk, which exudes a white latex when broken. None of these species has the bleach-like odor of **M. leptocephala** (p. 177).

Mycena haematopus

CAP: 1 to 3.5 cm in diameter, conical to bell shaped, surface smooth, dry, margin striate (especially in age), reddish or vinaceous brown, fading to pinkish brown, paler toward the margin, odor and taste mild. **GILLS:** Adnate or adnexed, close, pale pink, staining reddish brown, edges colored like the sides of the gills or red and contrasting with the paler faces (marginate). **STALK:** 3 to 7 cm long, 1 to 3 mm wide, equal, dry, dull reddish brown or paler, smooth or covered with scattered whitish fibrils, fragile, hollow, bleeding a reddish brown juice when cut, partial veil absent. **SPORES:** White, elliptical, smooth, and amyloid. **HABITAT:** Gregarious, in groups, or in tufts on well-decayed wood of conifers and, less commonly, hardwoods. **EDIBILITY:** Unknown but inconsequential in any case.

Mycena haematopus

The striate vinaceous-brown cap, bleeding stalk, and growth on wood distinguish this species. ***Mycena californiensis*** (*Mycena elegantula*) has a bleeding stalk, marginate gills (the edges reddish brown), and a reddish brown to orange-brown cap (p. 174). It grows in small groups or in extensive troops in leaf litter under oaks. ***Mycena purpureofusca*** (p. 182) has a pale purplish cap, gills edged in purple, a stalk that doesn't bleed, and a preference for rotting pine cones and decayed wood. In general, all of theses species have a long, thin stalk, striate cap, and white spores.

Mycena leptocephala

CAP: 1 to 2.5 cm in diameter, convex to bell shaped or sometimes slightly umbonate in age, smooth, translucent-striate, brown or grayish brown, fading to gray, odor of bleach when crushed. **GILLS:** Adnate, gray. **STALK:** 3 to 8 cm long, 1 to 2 mm wide, equal, fragile, hollow, smooth and polished to the naked eye, colored like cap or slightly darker, partial veil absent. **SPORES:** White, elliptical, smooth, amyloid. **HABITAT:** Scattered or in troops on ground under conifers and on wood chips, sometimes fruiting

Mycena leptocephala

in large numbers. **EDIBILITY:** Unknown but too small to be of value.

The distinctive bleach odor, brownish to gray, striate cap, and smooth fragile stalk typify this species. It occurs in the woods and on wood chips and woody debris in urban landscaped areas. Other species with a bleachlike odor (at least when crushed) include *M. stipata* (*M. alcalina*), which fruits on conifer wood, and *M. capillaripes,* which fruits in large troops on the ground under conifers during rainy periods. The gray gills of *M. capillaripes* are edged in pink (marginate), a characteristic not found in *M. stipata* or *M. leptocephala.* None of these species has a stalk that bleeds when cut (see *M. haematopus,* p. 176).

Mycena maculata

CAP: 2 to 4 cm in diameter, conical to bell shaped, becoming convex to plane, often with a low umbo, striate, dry, grayish brown fading to gray, often developing reddish brown spots in age, odor mild. **GILLS:** Notched to slightly decurrent, subdistant, and interspersed with veins, gray or tinted pink, often developing reddish spots. **STALK:** 4 to 8 cm long, 2 to 4 mm wide, equal, smooth, hollow, polished, grayish above, grayish brown below,

Mycena maculata

faintly striate, developing reddish brown spots, covered with dense white hairs at base, sometimes deeply rooted, partial veil absent. **SPORES:** White, elliptical, smooth, amyloid. **HABITAT:** Gregarious or commonly in clusters on conifer, and less frequently, hardwood logs and stumps. **EDIBILITY:** Unknown.

The reddish brown spots and stains that often develop on all parts of the fruiting body, the brownish color of the cap, and the growth on conifer wood, often in clusters, are hallmarks of this mushroom. The stalk does not bleed a reddish juice when cut, like **M. haematopus** (p. 176). Other species of *Mycena* that fruit on wood include **M. purpureofusca** (p. 182), which has a purplish cap and whitish gills edged in purple, and **M. galericulata**, which has a buff to pale tan cap, white to buff gills, and a mealy odor. It fruits scattered or in small clusters on rotting hardwood.

Mycena nivicola

(Other name: *Mycena griseoviridis*)

CAP: 1 to 3 cm in diameter, conical to convex to nearly flat, striate or sometimes slightly so, surface smooth, dark brown, olive

Mycena nivicola

brown, or tan, flesh yellow, odor mild or slightly mealy. **GILLS:**
Adnate, whitish or gray, becoming buff. **STALK:** 2 to 8 cm long, 2
to 3 mm wide, equal, viscid when moist, yellow above and reddish brown or mahogany brown below, partial veil absent.
SPORES: White, elliptical, smooth, amyloid. **HABITAT:** Single, scattered, or gregarious on the ground under conifers in the mountains, fruiting in spring after the snow melts, sometimes growing through the snow. **EDIBILITY:** Unknown.

Mycena nivicola (a provisional name) is a common snowbank mushroom has a dark cap, light-colored gills, and a colorful red and yellow stalk. It is easily recognized. *Mycena overholt-
sii,* another common snowbank mushroom, has a conical to
bell-shaped brown or grayish brown cap up to 6 cm in diameter,
which is relatively large for a *Mycena.* The stalk is 5 to 15 cm
long and up to 1.5 cm wide, colored like the cap but paler, and
the base is covered with white hairs. It typically grows in clusters
on conifer wood and stumps in spring near melting snow.

Mycena oregonensis

CAP: 2 to 8 mm in diameter, convex or bell shaped, bright yellow
to yellow-orange, fading in age, translucent-striate, odor and
taste mild. **GILLS:** Adnate to slightly decurrent, well spaced, yellow. **STALK:** 1 to 4 cm long, less than 1 mm wide, equal, fragile,
yellow or pale yellow, covered with a fine powder (pruinose),
inserted into the substrate, partial veil absent. **SPORES:** White,
teardrop shaped, smooth, inamyloid. **HABITAT:** Single to gregarious on decaying conifer needles and oak leaves. **EDIBILITY:** Unknown but much too small to be of interest.

This dainty mushroom is identified by its bright yellow cap
and thin yellow stalk. *Mycena acicula* is also diminutive, but it
has a bright red cap that fades to orange-red or orange and a
very thin, pruinose stalk that becomes smooth and orange- or
lemon-yellow in age. It grows on woody debris in the forest. *Mycena adscendens* (p. 419) occurs on wood and conifer cones. It
has a striate white cap up to 4 mm in diameter and a white stalk
with a disklike base attached to wood. All of these mushrooms
fruit during rainy periods and can be common in protected
areas, but they are easily overlooked.

Mycena oregonensis

Mycena pura

(Other name: *Prunulus purus*)

CAP: 1.5 to 5 cm in diameter, conical or convex, becoming plane or slightly umbonate, typically lilac or purple, aging grayish or pinkish lilac or paler, hygrophanous (fading as it loses moisture), smooth, striate, odor radishlike. **GILLS:** Adnexed, close, tinged lilac (but lighter than the color of the cap), the edges paler than the sides. **STALK:** 2 to 7 cm long, 3 to 7 mm wide, equal or with an enlarged base, smooth, hollow, fragile, colored like cap, partial veil absent. **SPORES:** White, elliptical, smooth, amyloid. **HABITAT:** Single or scattered or in small tufts on the ground in hardwood/coniferous forests. **EDIBILITY:** Unknown.

This species can be identified by the lilac color of the cap and stalk, lighter-colored gills, striate cap margin, and slight odor of radish. It is not infrequently encountered, but it does not grow in large numbers like some other *Mycena* species. Other common lilac or purplish mushrooms include **Clitocybe nuda** (p.

Mycena pura

151), a much more robust mushroom with pinkish spores, ***Laccaria amethysteo-occidentalis,*** which has a wider and distinctly fibrous stalk, and many species of *Cortinarius*, which are generally larger and have rusty brown spores and a cobweblike veil. ***Mycena purpureofusca*** (below) also has lilac tones, but it grows in clusters on woody debris.

Mycena purpureofusca

CAP: 0.5 to 2.5 cm in diameter, conical to bell shaped, smooth, dry but lubricious when moist, purple or vinaceous lilac with a paler or light brown margin, striate, odor and taste mild. **GILLS:** Adnate, moderately close, white to grayish with dark purplish red edges (marginate). **STALK:** 3 to 10 cm long, 1 to 3 mm wide, equal, hollow, fragile, polished, colored like cap but paler toward the apex, partial veil absent. **SPORES:** White, elliptical, smooth. **HABITAT:** In small groups or often in clusters on coniferous woody debris and rotting pine cones. **EDIBILITY:** Unknown.

This mushroom is characterized by its purplish or vinaceous, conical or bell-shaped cap, purplish edges of the gills, and penchant for growing on pine cones and coniferous wood. Reddish *Mycena* species include ***M. californiensis*** (p. 174), which has a red- to orange-brown cap, a stalk that bleeds red juice, and whit-

Mycena purpureofusca

ish gills with dark reddish brown edges. It grows in small groups or in extensive troops in leaf litter under oaks. ***Mycena haematopus*** (p. 176), which typically fruits on rotting wood in groups, has a reddish brown cap and bleeding stalk, a feature that readily separates it from *M. purpureofusca*.

Marasmiellus candidus

CAP: 0.5 to 3 cm in diameter, convex but soon plane, the center often depressed, dry, smooth but striate in age, white, often stained pink, flesh very thin and soft, odor mild. **GILLS:** Adnate to slightly current, very distant, of varying lengths, interspersed with veins, white, stained pink in age. **STALK:** 0.7 to 2 cm long, 1 to 3 mm wide, equal or tapered at either end, central or eccentric, white or grayish but becoming dark brown to black from the base upward in age, the base fibrillose, partial veil absent. **SPORES:** White, spindle to teardrop shaped, smooth, inamyloid. **HABITAT:** Gregarious on dead twigs and branches of dead conifers, hardwoods, and shrubs. **EDIBILITY:** Unknown but too small to be of value.

During wet weather, this small mushroom sometimes fruits abundantly on twigs on the forest floor. The white cap, which

Marasmiellus candidus

may develop pink stains in age, the very distant gills, and light-colored stalk that darkens in age from the base up are distinguishing characteristics. ***Marasmius calhouniae*** (p. 186) also has a white cap and stalk and widely spaced gills, but it fruits abundantly on the forest floor under conifers, not on wood.

Gymnopus dryophilus
(Other name: *Collybia dryophila*)

CAP: 2 to 6 cm in diameter, convex becoming plane, pale chestnut brown, light orange-brown, or tan fading to buff, the margin often paler and wavy in age, hygrophanous, the surface smooth and moist, odor and taste mild. **GILLS:** Adnexed to notched, crowded, white or buff. **STALK:** 2 to 6 cm long, 3 to 6 mm wide, equal, smooth, hollow, cartilaginous, colored like cap but paler at the apex, white mycelium adhering to the base, partial veil absent. **SPORES:** White to cream colored, elliptical, smooth, inamyloid. **HABITAT:** Scattered, gregarious, or clustered on ground in hardwood/coniferous forests, typically under oaks but widespread. **EDIBILITY:** Edible, but caution is necessary because this mushroom may be confused with less desirable, small, brownish species.

This species is common early in the mushroom season. ***Rhodocollybia butyracea*** (*Collybia butyracea*) is similar and also

Gymnopus dryophilus

occurs in many types of forests but is more commonly found under conifers. It has a smooth, lubricious, reddish brown cap. The gills are adnexed to free, close, white, with uneven or scalloped edges. The clavate stalk is up to 10 cm long and 1 cm wide, cartilaginous, hollow, smooth and whitish near the apex and reddish brown or reddish tan below. It also has white downy mycelium attached to the base of the stalk. The spores are cream colored or tinged pink.

Marasmius Species

The ability of the fruiting bodies to rehydrate in water and continue to produce spores after complete desiccation is the distinguishing feature of many species of this genus. The mushrooms are modestly sized with white spores and a ringless stalk. The stalks are tough and polished or wiry, sometimes like a bristle or horse hair. Most are saprobic on leaf litter and debris, but some grow on wood. They are generally too small to be of interest for the table, but *Marasmius oreades*, the Fairy Ring Mushroom, is a choice edible.

a. Growing in grass, especially lawns; cap creamy or buff............
....**Marasmius oreades**

Marasmius androsaceus

a. Growing in woods
  b. Stalk thin, stiff, bristlelike, reddish brown.............................
  ....**M. quercophilus**
  b. Stalk thin, stiff, bristlelike, black...
  ....**M. androsaceus** (see *M. quercophilus*)
  b. Stalk not bristlelike but may be thin
    c. Cap white and stalk white; on ground under conifers .....
    .... **M. calhouniae**
    c. Cap a shade of brown
      d. Odor garliclike; cap brown .... **M. copelandii**
      d. Odor mild; cap mahogany brown or vinaceous............
      .... **M. plicatulus**

Marasmius calhouniae

CAP: 1 to 3 cm in diameter, broadly convex to plane, the center often slightly depressed, white or faintly gray in age, smooth, the margin often wavy, fragile, thin, and translucent-striate when moist, odor mild. **GILLS:** Slightly decurrent, distant, white, intervenose (presence of veins between the gills). **STALK:** 2 to 4 cm long, 2 to 5 mm wide, more-or-less equal, white to light gray, often light brown at the base, fragile, smooth, hollow, partial veil absent. **SPORES:** White, smooth, elliptical. **HABITAT:** Scattered,

Marasmius calhouniae

gregarious, or in large troops on pine needles. **EDIBILITY:** Unknown.

This small mushroom sometimes fruits abundantly on the forest floor. It is especially common under pines along the central and northern California coast. The white translucent-striate cap, widely spaced decurrent gills, and white spores are distinguishing features. Other small all-white mushrooms that grow on the ground in the woods include ***Hygrocybe russocoriacea*** (p. 125), which has waxy gills, a relatively long stalk, an odor of cedar, and never fruits in large numbers like *M. calhouniae;* and ***Marasmiellus candidus*** (p. 183), which is similar to *M. calhouniae* but fruits on twigs and branches, has a white cap that is tinged pink, and has a gray or black stalk base.

Marasmius copelandii

(Other name: *Mycetinis copelandii*) Garlic Mushroom

CAP: 0.5 to 2 cm in diameter, convex to plane, deeply striate in age, pale brown, fading in age, dry, odor and taste of garlic. **GILLS:** Adnate or adnexed, close, cream colored to tan. **STALK:** 2 to 6 cm long, 1 to 3 mm wide, more-or-less equal, tough, hollow, reddish brown below, paler toward the apex, surface pubescent, partial veil absent. **SPORES:** White, teardrop shaped, smooth, in-

Marasmius copelandii

amyloid. **HABITAT:** Scattered to gregarious on leaf litter, especially on fallen oak, tanoak, and chinquapin leaves. **EDIBILITY:** Edible, used as a seasoning.

The strong garlic odor sets this mushroom apart from other small, light-colored mushrooms fruiting in forest leaf litter. The odor is sometimes evident before the mushrooms are spotted. The striate pale brown cap and minutely hairy stalk that is reddish brown below and paler brown above are also good field marks. **Marasmius scorodonius** (*Mycetinis scorodonius*) is similar but has a hairless stalk and prefers to fruit on dead twigs and stems instead of fallen leaves. Relatives include **M. androsaceus** (*Gymnopus androsaceus*) (p. 186) and **M. quercophilus** (p. 191), which grow on needles and leaves, have hairlike stalks, and lack a distinctive odor. **Micromphale sequoiae** and **M. arbuticola** grow on Coast Redwood debris or madrone bark, respectively. Both have small (approximately 1 cm broad) brownish caps and thin but not hairlike stalks; *M. arbuticola* has a garliclike odor.

Marasmius oreades
Fairy Ring Mushroom
CAP: 2 to 5 cm in diameter, convex to bell shaped, often with a broad umbo, cream colored, buff, or reddish tan, fading paler,

the surface smooth, dry, dried fruiting body reviving when moistened, odor and taste mild. **GILLS:** Adnexed to nearly free, broad and subdistant, of varying lengths, cream colored. **STALK:** 2 to 6 cm long, 2 to 6 mm wide, equal, smooth, cream colored to buff, darkening brown or reddish brown in age, tough, partial veil absent. **SPORES:** White, elliptical, smooth, inamyloid. **HABITAT:** In rings and arcs in grass, especially common in late spring through fall in irrigated lawns or year-round in warm areas. **EDIBILITY:** Edible and good, but the tough stalks are usually discarded.

This mushroom typically fruits in large numbers in fairy rings in lawns and pastures. The cap is cream colored or reddish tan and convex or umbonate with fairly well spaced, adnexed to free gills and the stalk is tough and pliant. Like other *Marasmius* species, dried specimens have the ability to revive when moistened. It might be confused with **Clitocybe dealbata** (p. 147), a poisonous lawn inhabitant with which it often grows, but that species is basically white and has slightly decurrent, closely spaced gills. It and many other nonedible mushrooms also grow in fairy rings.

Marasmius oreades

Marasmius
plicatulus

Marasmius plicatulus

CAP: 1 to 4 cm in diameter, conical, convex, or bell shaped, sometimes expanding to plane in age, mahogany brown or rusty brown or vinaceous, dry, velvety, striate or wrinkled in age, odor mild. **GILLS:** Adnate to nearly free, distant, whitish to buff, often tinged pink. **STALK:** 5 to 12 cm long, 2 to 3 mm wide, equal, brittle, smooth and polished, reddish brown at the base, paler above, with white mycelium often adhering to the base, partial veil absent. **SPORES:** White, elliptical, smooth, inamyloid. **HABITAT:** Scattered or gregarious on ground in leaf litter under a variety of trees in the woods. **EDIBILITY:** Unknown but too small to be of value.

The combination of the mahogany-colored cap, light-colored widely spaced gills, and shiny, thin, reddish brown stalk identifies this attractive mushroom. The ability to revive when moistened and the tough and polished stalk place it in the genus *Marasmius*. Other red mushrooms on the forest floor include the reddish waxy caps, such as ***Hygrocybe coccinea*** (p. 119) and

H. miniata (p. 120), which have bright red caps and stalks and waxy gills. *Mycena haematopus* (p. 176) is also reddish but is easily recognized by its growth on wood and a stalk that bleeds a reddish juice when cut.

Marasmius quercophilus

CAP: 2 to 5 mm in diameter, convex to nearly flat, the center often depressed, light brown to buff, sometimes paler at the margin, dry, striate and wrinkled, capable of reviving in water after drying, odor mild. **GILLS:** Adnexed, subdistant, cream colored, of various lengths. **STALK:** 1 to 3 cm long, less than 1 mm thick, equal, hairlike, smooth, tough and stiff, dark brown or reddish brown, sometimes with a paler apex, inserted into its substrate, black hairlike rhizomorphs (bundles of hyphae) usually visible at the base of the stalk, among the fruiting bodies, or on surrounding leaves, partial veil absent. **SPORES:** White, elliptical, smooth. **HABITAT:** Scattered or gregarious on leaves of hardwoods, especially oak and tanbark oak. **EDIBILITY:** Unknown but too small to be of value.

This is one of the horsehair fungi, so named for their thin, dark stalks. *Marasmius quercophilus* has a light brown cap and dark brown stalk and grows on individual fallen oak leaves.

Marasmius quercophilus

Marasmius androsaceus (*Gymnopus androsaceus*) (p. 186) is similar but has a brown to dark brown wrinkled cap and black stalk. Black rhizomorphs at the base of the stalk are usually noticeable. It prefers fallen needles of redwood and pine. *Marasmius pallidocephalus* grows on needles of Douglas-fir and hemlock. Once dry, all of these mushrooms revive in water.

Volvopluteus gloiocephalus

(Other names: *Volvariella gloiocephala, V. speciosa*)

CAP: 5 to 15 cm in diameter, oval or conical becoming convex to almost plane, sometimes umbonate, typically smooth but sometimes with a patch of universal veil tissue, pale gray to grayish brown, fading paler to almost white, viscid when moist, flesh white and soft, odor mild or radishlike. **GILLS:** Free, crowded, at first white and then becoming pink or reddish as the spores mature. **STALK:** 5 to 15 cm long, 1 to 2.5 cm wide, usually enlarged at the base, white or pale gray, smooth, dry, partial veil absent. **VOLVA:** Saclike, white, sometimes buried. **SPORES:** Salmon pink to pinkish brown, elliptical, smooth. **HABITAT:** Single or scattered in disturbed areas, lawns, gardens, cultivated soil, along roads, and so on. Typically fruiting in late winter and spring. **EDIBILITY:** Edible but watery and of poor quality.

Volvopluteus gloiocephalus is readily recognized by the smooth pale grayish cap, presence of a volva, absence of a ring, and salmon-colored spores. Before the spores mature, it can look like species of *Amanita*, but those are mycorrhizal with trees, whereas *V. gloiocephalus* is saprobic on organic matter in the soil. *Volvopluteus gloiocephalus* sometimes fruits prolifically in fallow cultivated fields. It is one of the most common mushrooms in inland urban areas of California.

Phyllotopsis nidulans

CAP: 2 to 10 cm broad, fan shaped, convex to nearly flat, pale orange to brown-orange, paler in age, covered with fuzzy hairs, the surface usually dry, odor and taste distinctly unpleasant. **GILLS:** Orange, close. **STALK:** Absent or very short, partial veil absent. **SPORES:** Pink or pinkish brown, elliptical, smooth, inamyloid. **HABITAT:** In clusters or tierlike overlapping shelves on rotting wood of both hardwoods and conifers. **EDIBILITY:** Inedible.

Volvopluteus gloiocephalus

The orange color of the fan-shaped cap, unpleasant odor, and growth on wood identify this distinctive mushroom. Several other fleshy gilled mushrooms with hairy caps and short or absent stalks are encountered on wood: **Lentinellus ursinus** has a brown fan-shaped cap with a brown pubescence and white spores with conspicuously serrated (toothed) gills that are diag-

Phyllotopsis nidulans

nostic; ***Crepidotus mollis*** (p. 231) has a flabby-textured, hairy, fan-shaped fruiting body and brown spores; ***Panus rudis*** (*Lentinus strigosus*) has a violet-tinged but soon reddish brown, conspicuously hairy, round or fan-shaped cap, a short eccentric, central, or lateral hairy stalk, and white spores and occurs on hardwood stumps and logs; and ***Panus conchatus*** is similar to *P. rudis* but with a minutely hairy or smooth violet-brown cap. The Oyster Mushroom, ***Pleurotus ostreatus*** (p. 132), is larger, and the cap surface is smooth.

Pluteus cervinus
Deer Mushroom

CAP: 3 to 12 cm in diameter, broadly convex to plane, sometimes with a low umbo, dark brown, brown, or grayish brown, surface smooth or faintly streaked with fibrils, moist in wet weather, odor and taste mild or radishlike. **GILLS:** Free, close, white becoming pink as the spores mature. **STALK:** 5 to 12 cm long, 0.5 to 2 cm wide, equal or wider at the base, white, smooth or faintly fibrillose, partial veil absent. **SPORES:** Pink to pinkish brown, elliptical, smooth. **HABITAT:** Single or scattered in small groups on wood or woody debris in forests and on wood chips in urban areas. **EDIBILITY:** Edible.

Pluteus cervinus

This mushroom is characterized by its brown-colored cap, pink spores, free gills, and growth on wood. Related species include **P. atromarginatus,** which has dark brown gill edges (i.e., marginate gills) and a streaked brown cap, and **P. romellii** (*P. lutescens*), a smaller mushroom (the cap is usually less than 5 cm in diameter) with a brown to yellowish brown cap and a yellow or greenish yellow stalk. Both of these species fruit on wood. The pink spores of *Pluteus* species might cause confusion with species of *Entoloma*, but those mushrooms grow on the ground. **Pluteus petasatus** is a robust relative that often grows in wood chips in landscaped areas. The cap is gray or grayish brown.

Entoloma bloxamii
(Other name: *Entoloma madidum*)

CAP: 5 to 15 cm in diameter, convex to plane, sometimes with a low umbo, tacky when moist, dark bluish gray or darker, streaked with fibrils, flesh white, odor mild or mealy. **GILLS:** Adnexed to notched, close, white to pale bluish gray becoming pink as the spores mature. **STALK:** 5 to 10 cm long, 1 to 3 cm wide, equal or slightly tapering toward the base, bluish gray near the apex, shading to white below, fibrillose, solid, firm, partial veil absent. **SPORES:** Salmon pink, round to oval, distinctly angular. **HABITAT:** Single or scattered on ground in hardwood/

Entoloma bloxamii

coniferous forests. **EDIBILITY:** Not recommended since some species of *Entoloma* are poisonous.

Entoloma bloxamii is easily recognized by the bluish gray to dark blue (almost black) cap, bluish gray tint of the otherwise whitish stalk, salmon-colored spores, and forest habitat. Some species of *Entoloma* in the subgroup *Leptonia* (see p. 200) are also bluish black but are very small relative to this substantial *Entoloma*. *Entoloma bloxamii* might be mistaken for a *Russula* species, but those mushrooms have a brittle texture (not fibrous like most mushrooms) and white to yellow spores. **Tricholoma griseoviolaceum** and relatives (p. 161) are similar to young specimens of *E. bloxamii*, but they have white spores.

Entoloma ferruginans

CAP: 3 to 12 cm in diameter, convex to plane or broadly umbonate, margin often wavy in age, brown to dark brown, smooth, odor bleachlike. **GILLS:** Adnexed to notched, whitish becoming pink as the spores mature. **STALK:** 4 to 12 cm long, 1 to 3 cm wide, equal, fibrillose, white with pale brown to brown tones, partial veil absent. **SPORES:** Salmon pink, round to oval, distinctly angular. **HABITAT:** Single or scattered on ground in oak woodlands, especially common in southern California but

Entoloma ferruginans

widespread. **EDIBILITY:** Not recommended since some species of *Entoloma* are poisonous.

The color of the cap of this mushroom is variable, but it is usually some shade of brown, although buff or yellowish caps also occur. The spores are salmon pink and angular under a microscope, and the mushroom has a bleachlike odor, especially when the tissue is crushed. Among the many look-alikes is *E. lividoalbum* (below), which has a tan, often umbonate cap, white stalk, and mild odor and may cause gastrointestinal upsets. Because these mushrooms are difficult to identify, large brown species of *Entoloma* should not be collected for the table.

Entoloma lividoalbum

CAP: 6 to 10 cm in diameter, convex to bell shaped becoming nearly flat with a broad umbo, the margin incurved when young, often wavy in age, smooth to faintly striate, the surface dry to moist but not viscid, yellow-brown to grayish brown, finely streaked, flesh white, not changing color when bruised, fragile, odor mild to mealy, taste mealy. **GILLS:** Adnate to notched, close, white, soon tinged with the color of spores. **STALK:** 7 to 12 cm long, 1 to 2.5 cm wide, equal, dry, finely streaked, white, bruising tan to brown, flesh white, solid, fi-

Entoloma lividoalbum

brous, partial veil absent. **SPORES:** Pinkish brown, angular. **HABI-TAT:** Single, scattered, to gregarious on ground under hard-woods, especially oaks, but occasionally found under conifers. **EDIBILITY:** Unknown but possibly poisonous; all *Entoloma* species should be avoided.

This species is commonly encountered under oaks. The cap is some shade of brown or yellowish brown, and the margin is often wavy at maturity. The mild to mealy odor separates it from **E. ferruginans** (p. 196), which has a nitrous or bleachlike odor and a cap that is often dark brown. Look-alikes occur under both hardwoods and conifers, but the nomenclature of many is unsettled. All of these may be poisonous and should be avoided.

Entoloma holoconiotum

(Other name: *Nolanea holoconiota*)

CAP: 2 to 6 cm in diameter, tan to brown to brownish orange, conical to bell shaped with a pointed umbo, the surface silky and fibrillose with faint striations, odor mild, taste mild to unpleasant. **GILLS:** Adnexed, subdistant, tan to light brown, soon becoming tinged with the color of spores. **STALK:** 3 to 7 cm long, 3 to 4 mm wide, colored like cap, striate, fragile and easily splitting lengthwise, white rhizomorphs often visible at the base of the stalk, partial veil absent. **SPORES:** Brownish pink, angular, smooth. **HABITAT:** Growing in groups or small clusters in spring under conifers in the mountains. **EDIBILITY:** Unknown.

As a group, species of *Entoloma* in the subgroup *Nolanea* can be identified by their pinkish spores, caps that are often conical and umbonate, and longitudinally streaked stalks that split easily upon handling. The name *Nolanea* is often retained by mushroom collectors because the unifying field characteristics justify an exclusive group for identification purposes. *Entoloma holoconiotum* is identified by a very pointed umbo, brown color, silky texture, and growth in spring in the mountains. **Entoloma propinquum** (*Nolanea proxima*) is common and widespread in the West. It has a brownish orange, hygrophanous (fading as it loses moisture), convex, rarely umbonate cap up to 4 cm broad and a mild to mealy odor. The margin of the cap is often paler than the center.

Entoloma holoconiotum

Entoloma sericeum

(Other name: *Nolanea sericea*)

CAP: 2 to 5 cm in diameter, convex becoming plane with or without a small umbo, dark brown fading light brown in age or when dry, the surface of the cap glossy and silky, especially when dry, odor and taste mealy. **GILLS:** Adnate to notched, light brown, soon becoming tinged with the color of spores. **STALK:** 3 to 5 cm long, 3 to 8 mm wide, colored like the cap but the base whitish, fragile and easily splitting lengthwise, partial veil absent. **SPORES:** Salmon pink, angular. **HABITAT:** Growing in groups in lawns, pastures, disturbed areas, and woods, often associated with grasses. **EDIBILITY:** Unknown.

This small mushroom is common in grassy areas in many habitats, including turf. The pink spores and longitudinally striate stalk that splits easily are distinctive characteristics of the subgroup of *Entoloma* species previously known as the genus *Nolanea*. The silky cap that may not retain a prominent umbo helps identify *E. sericeum*, which represents a complex of species in the West, none of which is likely the true *E. sericeum* of Europe. **Entoloma strictius** (*Nolanea stricta*) has a conical, umbonate, grayish brown cap, a long stalk, and mild odor. It occurs

Entoloma sericeum

on the ground in moist woods or on rotting wood. ***Entoloma hirtipes*** (*Nolanea hirtipes*) has a brown silky cap, long stalk, and an unpleasant odor. None of these species should be considered edible.

Entoloma parvum
(Other name: *Leptonia parva*)

CAP: 1 to 3 cm in diameter, convex, often becoming shallowly depressed in age, appressed fibrillose, dark blackish blue (appearing almost black in the deep woods), aging grayish brown with a blue tinge, odor mild. **GILLS:** Adnate to notched, whitish but soon pink as the spores mature. **STALK:** 1 to 6 cm long, 2 to 4 mm wide at the apex, equal, smooth, bluish gray to blackish blue, partial veil absent. **SPORES:** Pink, angular. **HABITAT:** Scattered on the ground in hardwood/coniferous forests, especially under Coast Redwoods, commonly encountered but not in quantity. **EDIBILITY:** Unknown but too small for the table.

As a group, *Entoloma* species in the subgroup *Leptonia* are small saprobic mushrooms of forests that parade in various shades of black, blue, or violet. Some are iridescent. Common relatives include ***E. chalybeum*** (*Leptonia chalybaea*) which has a

Entoloma parvum

dark bluish black cap, bluish black stalk slightly paler than the cap, and bluish gray gills and occurs on the litter of coastal redwood and other conifers; and ***E. serrulatum*** (*Leptonia serrulata*), found in leaf and twig litter under a variety of trees, which has a dark blackish blue cap and stalk and bluish gray gills that have blackish, minutely serrulate (jagged) edges. Many other similar species are easily overlooked in the dim light of the forest. ***Entoloma sericellum*** (*Alboleptonia sericella*) also has angular pink spores, but this small fragile mushroom is completely white. The cap is broadly convex and up to 4 cm in diameter; the stalk is up to 5 cm long and less than 8 mm wide. It grows on the ground in hardwood/coniferous forests.

Chroogomphus vinicolor
Pine Spike

CAP: 3 to 9 cm in diameter, conical with an inrolled margin at first, becoming rounded-conical to convex or umbonate, vinaceous brown, mahogany brown, or reddish brown, viscid when moist, silky and shiny when dry, smooth or appressed fibrillose, flesh orange or vinaceous orange, taste and odor mild. **GILLS:** Decurrent, fairly well spaced, orange becoming gray or black as the spores mature. **STALK:** 3 to 12 cm long, 0.5 to 2 cm wide, often curved, tapered toward the base, dry, solid, smooth or

Chroogomphus vinicolor

with scattered fibrils, brownish orange becoming vinaceous in age, partial veil yellowish to orange, disappearing altogether or leaving a zone of hairs high on the stalk. **SPORES:** Black, elliptical, smooth. **HABITAT:** Single or more often scattered under conifers, especially pines, where it parasitizes mycelium of *Suillus* and *Rhizopogon* species. **EDIBILITY:** Edible.

The reddish, orange, or reddish orange color of all parts of the fruiting body, decurrent gills, black spores, and brownish orange stalk are a distinctive set of features of the Pine Spike. In dry weather, it has a clean silky appearance. The above description also fits ***C. ochraceus***, but that species has ochre or olive hues in the cap and has thin-walled cystidia (whereas *C. vinicolor* has thick-walled cystidia). ***Chroogomphus pseudovinicolor*** is a larger relative that has a dry dull red or orange cap, thick stalk up to 5 cm wide covered with reddish fibrils, and olive black spores. ***Chroogomphus tomentosus*** has an ochre-colored cap that develops purplish colors in age. The surface of the cap is dry, fibrillose, and wooly.

Gomphidius glutinosus

CAP: 3 to 10 cm in diameter, convex becoming plane or depressed in the center, brownish or purplish gray or pale reddish

Gomphidius glutinosus

brown, often stained black in spots and blotches in age, viscid to slimy when moist, smooth, flesh white or gray, odor mild. **GILLS:** Decurrent, thick and waxy, fairly well spaced, whitish at first and then black as the spores mature. **STALK:** 5 to 10 cm long, 1 to 2 cm wide, equal or tapered slightly toward the base, viscid to slimy when moist, white above and bright yellow at the base, bruising or staining black in age, partial veil leaving a faint slimy ring on the stalk. **SPORES:** Black, elliptical, smooth, 15 to 21 by 4 to 8 μm. **HABITAT:** Single or scattered on ground under conifers, especially common under Douglas-fir. **EDIBILITY:** Edible but very slimy.

The slimy cap, decurrent gills, black spores, and bright yellow color of the lower stalk distinguish this common mushroom. Like other species of *Gomphidius*, it is always associated with conifers. **Gomphidius oregonensis** is similar but tends to grow in clusters. It has smaller spores (11 to 14 by 5 to 6 μm) and a stalk that is often deeply rooted. **Gomphidius subroseus** has a light red or rose-colored small cap (up to 6 cm broad). It also has a white stalk with a yellow base.

Inky Caps

Most inky caps are united by a common feature: the caps, gills, or both deliquesce (liquefy) from the margin of the cap toward the center as the spores mature. The species included here are not necessarily closely related, although they have been historically grouped together. They have black spores, and all grow as saprobes on decaying plant material and dung.

a. Cap up to 15 cm or more tall, cylindrical, white ....
....**Coprinus comatus**

a. Cap not cylindrical, shorter, or both
  b. Cap more than 5 cm broad, silky, gray ....
  ....**Coprinopsis atramentaria**
  b. Cap less than 5 cm broad, variously colored
    c. Cap less than 1 cm broad, whitish ....
    .... **Coprinopsis stercorea**
    c. Cap more than 1 cm broad
      d. Cap gray; cap and gills completely deliquescent...........
      ....**Coprinopsis lagopus**

Coprinellus flocculosus

  d. Cap whitish, buff, or brown, partially or not deliques-
     cent
     e. Cap grayish or yellow-brown, non- or weakly deli-
        quescent ....***Pseudocoprinus lacteus***
     e. Cap dull yellow-brown, gills partially deliquescent
        f. Growing in clusters at the base of trees ....
            .... ***Coprinellus micaceus***
        f. Growing scattered in wood chips and woody
            debris..
            .... ***Coprinellus flocculosus*** (see *C. micaceus*)

Coprinellus micaceus
(Other name: *Coprinus micaceus*)
CAP: 2 to 5 cm in diameter, oval becoming bell shaped to convex,
yellowish brown, striate nearly to center, surface covered with
sparkling granules that may wash off, leaving the cap bald, in
age graying and deliquescing toward the center, the margin becoming tattered. **GILLS:** Adnexed to notched, close, pinkish gray
and finally black, deliquescing but not always entirely. **STALK:** 3
to 6 cm long, 3 to 5 mm wide, white, smooth, hollow, partial veil

Coprinellus micaceus

absent. **SPORES:** Black, elliptical, smooth. **HABITAT:** Clustered or scattered on woody debris; often encountered in clusters at the base of trees or stumps. **EDIBILITY:** Edible.

Formerly placed in the genus *Coprinus*, this mushroom is recognized by the shiny granules on the surface of the yellowish brown caps, tendency to fruit in clusters, and gills that entirely or partially deliquesce. In California, it fruits any time of the year with moisture and is fairly common in summer around the base of trees in irrigated lawns. Because the cap granules quickly wear away, this species might be mistaken for **Coprinopsis atramentaria** (below), a larger inky cap that also fruits in clusters on woody debris. It typically has a silky gray cap with a much more substantial stalk. **Coprinellus flocculosus** (p. 204) has dingy yellow caps with thick white or translucent veil fragments. It grows scattered or in small groups in wood chips in landscaped areas. The base of its stalk may have a small ring, at least when young.

Coprinopsis atramentaria

(Other name: *Coprinus atramentarius*) Inky Cap

CAP: 5 to 10 cm in diameter, oval at first becoming bell-shaped, gray to brownish gray, silky, dry, smooth or with a few small scales on the center, margin striate or pleated, deliquescing in age toward the center. **GILLS:** Free, crowded, white becoming pinkish gray and finally black. **STALK:** 5 to 15 cm long, 1 to 2 cm wide, white and silky, equal or sometimes with an enlarged base, dry, hollow, partial veil fibrous, white, leaving a basal ring that usually disappears. **SPORES:** Black, elliptical, smooth. **HABITAT:** Scattered or more typically clustered on woody debris or near buried wood, especially common where a tree has been removed; fruiting year-round when moisture is available. **EDIBILITY:** Edible with caution; this mushroom should not be consumed with alcohol.

The lead gray silky cap, deliquescing gills, and occurrence in clusters are the primary field marks. The common name Inky Cap refers to the autodigestion (liquefying) nature of the gills as the spores mature from the bottom of the gills toward the center of the cap. In contrast to most mushrooms, the spores are not efficiently released into air currents. Instead, they are released as the gills become digested and exposed. The resulting inky mass can be used as writing ink.

Coprinopsis atramentaria

Coprinopsis lagopus
(Other name: *Coprinus lagopus*)

CAP: 2 to 5 cm in diameter, oval to conical when young but soon expanding to plane with an upturned margin, often split and tattered in age, surface at first covered with white to gray hairs that quickly wear away, revealing a gray to very dark gray surface, conspicuously striate at maturity. **GILLS:** Free, close, gray becoming black and inky, quickly deliquescing in moist weather. **STALK:** 5 to 12 cm long, 2 to 5 mm wide, white, equal, hollow, fragile, covered with white fibrils from universal veil remnants, partial veil absent or if present, quickly disappearing. **SPORES:** Black, elliptical, smooth. **HABITAT:** Scattered on woody debris or leaf litter, sometimes in great numbers. **EDIBILITY:** Unknown but too insubstantial to be of value.

This common inky cap is characterized by the dense white hairs that cover young fruiting bodies and the gray striate cap that curls back on itself in maturity and quickly deliquesces in moist weather. The cap often disappears in just a few hours, leaving an upright stalk that doesn't last much longer. It is not uncommon to find hundreds of specimens fruiting on ground among woody debris during rainy weather.

Coprinopsis lagopus

Coprinopsis stercorea
(Other name: *Coprinus stercoreus*)
CAP: 5 to 10 mm in diameter, oval becoming convex to plane, white or light gray, deeply striate or pleated, very delicate and fragile, covered with white granules, the margin hairy at least when young. **GILLS:** Adnexed, whitish becoming black, fragile, shriveling or partially deliquescent. **STALK:** 1 to 4 cm long, 1 to 2 mm wide, white, equal, fragile, hollow, covered with white hairs when young but often smooth in age, partial veil absent. **SPORES:** Black, elliptical, smooth. **HABITAT:** On dung of horses, cows, and other animals, and in compost piles, fruiting during periods of very high humidity, especially overcast or foggy periods after rain. **EDIBILITY:** Unknown; too small to be of value.

This is one of the many mushrooms that fruit on dung of herbivores and on compost piles during periods of rain. Because the fruiting body quickly disappears in the full sun, it is often only encountered in the morning hours. When young, this mushroom is easily recognized by the white sugary granules on the cap surface and hairs on the margin. **Coprinopsis ephemeroides** grows on dung but has a ring. Other dung-loving mushrooms include several species in the genera *Panaeolus* and *Psilocybe*, among others.

Coprinopsis stercorea

Coprinus comatus
Shaggy Mane

CAP: 4 to 8 cm in diameter, 6 to 20 cm tall, cylindrical to oval, white, sometimes with a pale brown center, covered with white and brown flat or recurved scales, dry, deliquescing and blackening from margin to the center. **GILLS:** Free, very crowded, white and then pink, finally becoming black and inky. **STALK:** 6 to 20 cm long, 1 to 2 cm wide, equal or with a bulbous base, white, hollow or stuffed with fibrils, partial veil leaving a ring that slides to the base of the stalk or sometimes disappears. **SPORES:** Black, elliptical, smooth. **HABITAT:** Single, scattered, or in clusters in grass, along roads, and on disturbed soil. **EDIBILITY:** Edible and choice.

Because the Shaggy Mane is not likely to be confused with other mushrooms, it is highly sought after for the table. The shaggy, tall cylindrical cap that dissolves into ink at maturity is distinctive. It sometimes fruits in great numbers in grassy areas and disturbed soils but at times curiously pushes up through asphalt. ***Podaxis pistillaris***, a conspicuous and common desert and chaparral fungus, resembles a dried Shaggy Mane but lacks

Coprinus comatus

gills. Its fruiting bodies dry in place and continue to release spores for years. In other parts of the world it is associated with termite mounds; in American deserts it might be associated with ant colonies and the plant material they store underground.

Pseudocoprinus lacteus

(Other names: *Coprinus plicatilis, Parasola leiocephala*)

CAP: 1 to 3 cm in diameter, oval becoming convex to plane, pale yellow-brown or grayish brown, the center brown, deeply and conspicuously striate and pleated nearly to the center, very fragile and ephemeral. **GILLS:** Free, attached to a collar around the stalk, gray, fragile, not deliquescing. **STALK:** 2 to 6 cm long, 1 to 2 mm wide, equal, smooth, fragile, white or off-white, partial veil absent. **SPORES:** Black, elliptical, smooth. **HABITAT:** Scattered or gregarious on grass thatch, decaying wood, compost piles, and disturbed ground during periods of high moisture. **EDIBILITY:** Too small to be of any value.

Pseudocoprinus lacteus is recognized by a yellow-brown, delicate, pleated cap with a brown center, the attachment of the gills to a collar around the stalk, and the nondeliquescing nature of the fruiting body. It is common during periods of very

Pseudocoprinus lacteus

high humidity, especially overcast or foggy periods after rain. **Parasola auricoma** (*Coprinus auricomus*) is similar, but the gills are typically attached to the stalk and weakly deliquescent and the striate cap is grayish brown with a pale orange-brown center. It grows in lawns or on woody debris. **Coprinellus disseminatus** (*Coprinus disseminatus*) has a whitish to grayish brown, deeply striate cap about 1 cm broad. The gills do not deliquesce. It fruits in large groups on buried woody debris or at the base of trees.

Panaeolus foenisecii

(Other name: *Panaeolina foenisecii*) Haymaker's Mushroom
CAP: 1 to 3 cm in diameter, conical to convex, dull brown to chestnut brown, hygrophanous (fading as it loses moisture), fading to tan (but sometimes retaining a darker band near the margin), smooth, fragile, flesh brownish, odor and taste mild.
GILLS: Adnate or pulling away from the stalk, close, brown becoming dark brown, the edges lighter in color, the sides mottled as the spores mature. **STALK:** 3 to 8 cm long, 1 to 3 mm wide, equal, hollow, buff to tan, brown at the base, partial veil absent.
SPORES: Dark brown or purple-brown, elliptical, roughened by minute warts. **HABITAT:** Scattered or gregarious in lawns and other grassy areas during warm weather. **EDIBILITY:** To be

Panaeolus foenisecii

avoided, possibly containing small amounts of the hallucinogenic compounds psilocybin and psilocin.

The hygrophanous, brown, conical to convex cap, fragile hollow stalk, and gills that become mottled as the spores mature in groups are the principal field marks of this common lawn inhabitant. It fruits in the warmer times of the year, sometimes in large numbers, with other lawn mushrooms, such as **Marasmius oreades** (p. 188), **Agaricus campestris** (p. 224), and **Conocybe apala apala** (p. 249). Compare *P. foenisecii* with **Psathyrella candolleana** (p. 233) and **P. gracilis** (p. 234), which may also fruit in or near lawns. They have white stalks and lack the mottled gills of *P. foenisecii.*

Panaeolus solidipes

CAP: 3 to 8 cm in diameter, rounded bell shaped, white or white with pale silver or gray tones, surface dry, smooth or wrinkled, flesh white, odor mild. **GILLS:** Adnate to adnexed, close, the edges white, the sides gray becoming mottled with darker areas from maturing spores, blackish in age. **STALK:** 6 to 15 cm long, 0.5 to 1.5 cm wide, equal or wider at the base, solid, smooth or longitudinally lined, white or colored like cap, beaded with water when fresh, partial veil absent. **SPORES:** Black, elliptical,

Panaeolus solidipes

smooth. **HABITAT:** Single, scattered, or gregarious on dung of horses and other livestock or on compost. **EDIBILITY:** Edible.

This relatively large *Panaeolus* species can be recognized by its whitish cap, ringless, white solid stalk that is often beaded with water, black spores, and growth on dung. ***Panaeolus papilionaceus*** (*P. campanulatus, P. retirugis*) is another common dung-inhabiting species characterized by a grayish or reddish brown cap 1 to 4 cm broad and edged with veil tissue, mottled gills, black spores, and a slender stalk. ***Panaeolus semiovatus,*** which also fruits on dung, has a viscid, buff-colored cap and a ring that collapses on the stalk. The mottled nature of the gills is due to the maturation of spores in clusters rather than uniform maturation like most mushrooms.

Psilocybe semilanceata
Liberty Cap
CAP: 1 to 3 cm in diameter, conical or bell shaped with a pointed umbo, olive brown to brown aging buff, smooth, viscid, the margin often blue. **GILLS:** Adnate to adnexed, gray becoming purplish as spores mature. **STALK:** 4 to 15 cm long, 1 to 3 mm wide, equal, white to buff, tan, occasionally bruising and aging blue, partial veil usually absent. **SPORES:** Purple-brown, ellipti-

cal, smooth. **HABITAT:** Scattered to gregarious in pastures and grass in the Pacific Northwest. **EDIBILITY:** Hallucinogenic.

The purple-brown spores, pointed olive brown cap, long stalk, and occurrence in grass help identify this species. The bluing reaction, which indicates the presence of the hallucinogenic compound psilocybin, occurs faintly or not at all in this species (other mushrooms bruise blue for other, unknown reasons). ***Psilocybe cubensis*** (p. 424) is the Magic Mushroom cultivated and used recreationally. It has a tan viscid cap, fragile ring on the stalk, and readily stains blue when bruised. It is most commonly found on dung in tropical areas but occasionally turns up on the West Coast on decaying wood chips. ***Psilocybe cyanescens*** is more commonly encountered on wood chips; it has a chestnut brown to buff, hygrophanous cap with a wavy, striate margin and a low umbo, and an ephemeral ring on the stalk. The brown-capped ***Deconica coprophila*** (*P. coprophila*), which does not stain blue, is common on dung. It has a striate cap margin edged with white veil tissue.

Psilocybe semilanceata

Hypholoma fasciculare

Hypholoma fasciculare
(Other name: *Naematoloma fasciculare*) Sulfur Tuft
CAP: 2 to 5 cm in diameter, convex becoming slightly umbonate or plane, yellow or yellow-orange, moist, smooth, flesh yellow, odor mild, taste bitter. **GILLS:** Adnate, close, yellow becoming greenish yellow and finally purple-brown from spores. **STALK:** 3 to 10 cm long, 4 to 10 mm wide, more-or-less equal, hollow, dry, firm, yellow developing brownish orange stains in age or where bruised, partial veil evanescent, leaving thin remnants on the margin of the cap or a fibrillose ring on the stalk that soon disappears. **SPORES:** Purple-brown, elliptical, smooth. **HABITAT:** Scattered, gregarious, or often in dense clusters on decaying hardwood or conifer logs or stumps. **EDIBILITY:** Poisonous.

The large, bright yellow clusters of fruiting bodies of this species on fallen hardwood or conifer logs, stumps, or buried wood are a conspicuous part of the mushroom flora in western forests. The purple-brown spores are often readily visible on caps dusted by spores from caps above them. The greenish yellow gills and bitter taste separate it from its close relative, ***H. capnoides*** (*Naematoloma capnoides*), which has a yellow to reddish brown cap, grayish gills, and a mild taste. It also grows in clusters but is limited to decaying coniferous wood. Unlike *H. fas-*

ciculare, *H. capnoides* is edible, but the two species are sometimes difficult to differentiate.

Stropharia ambigua

CAP: 5 to 15 cm in diameter, convex to plane, yellow to buff, the margin often edged with white remnants of the partial veil, viscid when moist, smooth. **GILLS:** Adnate, close, gray becoming purplish black as the spores mature. **STALK:** 6 to 18 cm long, 1 to 2 cm wide, equal, white, smooth or silky above the ring, covered with soft white scales below the ring, partial veil leaving remnants on the cap margin and a fragile evanescent ring on the stalk; white rhizomorphs often adhere to the base of the stalk. **SPORES:** Purple-black, elliptical, smooth. **HABITAT:** Single, scattered, or gregarious on ground in hardwood/coniferous forests. **EDIBILITY:** Unknown.

The yellow cap, veil remnants that hang on the margin of the cap, scaly stalk, and purple-black spores distinguish this attractive mushroom. Other *Stropharia* species with viscid and modestly sized caps (less than about 5 cm broad) include **S. coronilla,** which has a yellowish cap, dry, short stalk with a striate ring, and a preference for grassy areas, and **S. semiglobata,** which has a yellowish cap, viscid stalk with a fibrillose ring, and a habitat of dung, compost, and well-manured lawns. **Stropharia aerugi-**

Stropharia ambigua

nosa (p. 425) is easily identified by its bluish green cap. It grows in a variety of habitats, including landscaped areas.

Leratiomyces percevalii
(Other name: *Stropharia riparia*)

CAP: 2 to 10 cm in diameter, hemispherical becoming broadly convex to nearly plane, yellow to cream colored, lubricious but soon dry, smooth or with scattered ochre to tan scales, when young the margin often edged with white remnants of the partial veil, odor and taste mild. **GILLS:** Adnate to slightly decurrent, close, whitish when young, becoming purplish gray and mottled in age as the spores mature. **STALK:** 4 to 13 cm long, 0.5 to 1.5 cm wide, more-or-less equal, sometimes twisted, white or cream colored, dry, the base often with adhering mycelia, partial veil leaving a fibrillose ring that darkens from purplish spores or may disappear in age. **SPORES:** Purple-brown, elliptical, smooth. **HABITAT:** Scattered to gregarious in grassy areas, wood chips, and near trails. **EDIBILITY:** Unknown.

The yellowish, broadly convex cap edged with veil remnants when young, whitish stalk that is long and twisted in some specimens, fibrillose ring often dusted with purplish spores, and occurrence in grassy places and wood chips are useful characteristics for field identification. It sometimes fruits in large numbers. The size of the caps can be highly variable. In poor soils, sizes of

Leratiomyces percevalii

caps and numbers of fruiting bodies are diminished relative to growth and extent of fruiting in fresh and nutritionally rich wood chips. Other yellow or buff species found in the same habitats include **Psathyrella candolleana** (p. 233), **P. gracilis** (p. 234), **Agrocybe putaminum** (p. 239), and **A. praecox** (p. 239), among others.

Leratiomyces ceres

(Other names: *Hypholoma aurantiacum, Stropharia aurantiaca*)

CAP: 2 to 6 cm in diameter, convex becoming plane or broadly umbonate, bright reddish brown or reddish orange, tacky when moist, otherwise dry, smooth, the margin often edged with white veil fragments, odor and taste mild. **GILLS:** Adnate or notched, light grayish coming brown and then purple-brown as spores mature. **STALK:** 3 to 6 cm long, 5 to 10 mm wide, equal or enlarged at base, dry, whitish staining reddish orange in age, partial veil leaving a thin white ring that typically disappears; white or yellow mycelium often adhering to the base of the stalk. **SPORES:** Dark purple-brown, elliptical, smooth. **HABITAT:** Gregarious on wood chips or lawn thatch in landscaped areas. **EDIBILITY:** Unknown.

Leratiomyces ceres is recognized by a bright reddish cap, purple-brown spore print, and preference for wood chips in

Leratiomyces ceres

landscaped areas. Many other mushrooms are common in wood chips, including **Psathyrella gracilis** (p. 234), which has a brown hygrophanous cap and fragile thin stalk, **Agrocybe putaminum** with brown spores and caps that are often superficially cracked, **Gymnopilus sapineus** (p. 246), an orange-capped species with a rusty orange spore print, and the modestly sized **Tubaria furfuracea** (p. 251), which has an orange-brown spore print, among many others.

Agaricus Species

The commercial Button Mushroom, *Agaricus bisporus*, is the most familiar species of this genus. Some relatives collected in landscaped areas and the woods are equally edible, but some are poisonous, causing severe gastrointestinal upsets. As a group they are typically medium to large in size, dull white to brownish, and often stain yellow, brown, or red when injured. The caps are convex or flattened at maturity and smooth or covered by fibrils or scales. A ring on the stalk is usually present and conspicuous. The gills are free from the stalk and white, gray, to pink when young, later becoming dark brown. The odor, which aids identification of some species, ranges from mild to almond-extract-like to phenolic (unpleasant and resinlike). Spores are chocolate brown. These mushrooms live on decaying plant matter such as lawn thatch, manure, and compost.

a. Odor at very base of stalk phenolic when crushed
    b. Base of stalk quickly yellowing when cut, cap white and smooth ....**Agaricus xanthodermus**
    b. Base of stalk not bruising bright yellow (if bruising pale yellow, then cap scaly)
        c. Cap smooth, white; widespread ....**A. californicus**
        c. Cap with small appressed fibrils; in or near woods
            d. Cap fibrils pinkish brown ....**A. hondensis**
            d. Cap fibrils gray ....**A. moelleri** (see *A. hondensis*)

a. Odor at very base of stalk mild or sweet
    e. Cap white, smooth or slightly scaly (but may become scaly in dry weather)
        f. Ring poorly defined, in grass ....**A. campestris**
        f. Ring membranous

> g. In grassy areas; cap yellowing when bruised; odor sweet ....**A. arvensis**
>
> g. In woods; cap yellowing when bruised or in age; odor sweet ....**A. silvicola**
>
> g. In hard-packed soil, often urban; odor mild.................
> ....**A. bitorquis**

e. Cap white with yellowish or brownish fibrils or scales

> h. Odor almondlike; cap yellow-fibrillose, staining yellow..
> ....**A. augustus**
>
> h. Odor not almondlike; cap whitish, colored with fibrils, not staining yellow
>
> > i. Cap covered with brown fibrils; in fields and disturbed soils ....**A. cupreobrunneus**
> >
> > i. Cap covered with purple-brown fibrils; in woods........
> > ....**A. subrutilescens**

Agaricus arvensis

Horse Mushroom

CAP: 7 to 20 cm in diameter, oval becoming broadly convex, white to creamy, smooth or slightly scaly, dry, bruising yellow at least when young, odor of almond or anise. **GILLS:** Free, white at first becoming dark brown in age. **STALK:** 5 to 12 cm long, 1 to 3 cm wide, white, sometimes staining or aging yellowish, equal or with an enlarged base, smooth or slightly scaly near the base, the partial veil leaving a membranous, skirtlike white ring with cottony patches on the lower surface. **SPORES:** Chocolate brown, elliptical, smooth. **HABITAT:** Single or scattered or in groups in fairy rings in meadows, fields, lawns, pastures, and other grassy areas. **EDIBILITY:** Edible.

This robust, widespread grass inhabitant is characterized by a smooth, dry whitish cap that usually stains yellow when bruised, sweet odor, free gills, chocolate brown spores, and skirtlike ring with cotton patches on its underside. Rub the margin of the cap to check for the yellowing reaction. *Agaricus arvensis* can be confused with the toxic **A. xanthodermus** (p. 230), which may grow in the same habitats. That mushroom has a phenolic odor (unpleasant, resinlike), and the flesh of the base of the stalk bruises yellow quickly; in contrast, the base of the stalk of *A. arvensis* does not stain yellow or slowly bruises yellow.

Agaricus arvensis

Agaricus augustus
The Prince

CAP: 8 to 30 cm in diameter, marshmallow shaped to convex becoming plane in age, whitish to yellow with yellowish brown or brown fibrils, aging and staining yellow, eventually developing a golden hue, odor fragrant (like almond paste). **GILLS:** Free, whitish at first and then pink becoming chocolate brown. **STALK:** 7 to 30 cm long, 2 to 6 cm wide, white but aging or bruising yellow, equal or enlarged at the base, smooth above the membranous, persistent skirtlike ring, shaggy with white scales below the ring, at least when young. **SPORES:** Chocolate brown, elliptical, smooth. **HABITAT:** Single or often in clusters in woods and gardens, especially in disturbed soil along paths, roads, and such, fruiting much of the year during mild weather. **EDIBILITY:** Edible and choice.

The large size, yellow-staining cap, chocolate brown gills and spores, partially shaggy stalk, and odor of almond paste distinguish this prized edible mushroom. The golden color that devel-

Agaricus augustus

ops as the fruiting body ages is also distinctive. **Agaricus moelleri** (*A. praeclaresquamosus*) is also large, similarly shaped, and grows in the same habitats (p. 424). In contrast to *A. augustus*, it has grayish brown cap fibrils, an unpleasant odor, a smooth stalk both above and below the ring, and stains vinaceous brown when bruised. It is toxic to many and should be avoided. The name *A. moelleri* is borrowed from a European collection; undoubtedly the species in the West is unique.

Agaricus bitorquis

CAP: 5 to 15 cm in diameter, convex to plane or slightly depressed, whitish but often dirty white from adhering soil, smooth but scaly in dry weather, the surface dry, flesh firm and unchanging when bruised, odor and taste mild. **GILLS:** Free, grayish pink becoming dark brown. **STALK:** 3 to 8 cm long, 1 to 3 cm wide, white, firm, equal or enlarged at the base, partial veil white and membranous, leaving a thick, persistent ring on lower stalk with flaring upper edge. **SPORES:** Chocolate brown, elliptical, smooth. **HABITAT:** Single or scattered in disturbed areas, especially hard-packed soil, often fruiting underground with just the top of the cap visible. **EDIBILITY:** Edible and choice.

The white, smooth, and firm cap, free gills, chocolate brown spores, short and compact stature, and mild odor define this

Agaricus bitorquis

mushroom of hard-packed soil. If one specimen is found, the surrounding area should be carefully examined since fruiting bodies often barely break the surface of the soil. It might be confused with **A. bernardii**, but that species has a mild or briny odor and stains reddish when bruised, and the surface of the cap often breaks up into scales or warts. Despite its briny odor, *A. bernardii* is a good edible.

Agaricus californicus

CAP: 5 to 12 cm in diameter, marshmallow shaped to convex and finally plane, white or sometimes white with a pale grayish brown center or brownish from a layer of brown fibrils, dry, flesh not discoloring when bruised or faintly yellowing, odor faintly of phenol (unpleasant and resinlike). **GILLS:** Free, whitish at first and then pink and finally chocolate brown. **STALK:** 3 to 10 cm long, 1 to 2 cm wide, white, smooth or slightly fibrillose, equal or slightly enlarged at base, partial veil leaving a thick, membranous, white, persistent ring on upper stalk. **SPORES:** Chocolate brown, elliptical, smooth. **HABITAT:** Scattered or gregarious in lawns, parks, and gardens. **EDIBILITY:** Toxic, causing gastric upsets.

This mushroom occurs in a wide variety of habitats. Because it is mildly toxic and causes confusion with the edible and popu-

Agaricus californicus

lar **A. campestris** (below), it is a good mushroom to know. The defining characteristics are the whitish gills and marshmallow-shaped cap of young specimens, the thick persistent ring, and odor of phenol, which may be faint in some specimens. In contrast, *A. campestris* has a mild odor and weakly developed ring. **Agaricus xanthodermus** (p. 230) is similar, but it quickly stains yellow upon bruising. **Agaricus arvensis** (p. 220) occurs in lawns and has a sweet odor, membranous ring, and faint yellow reaction to bruising.

Agaricus campestris
Meadow Mushroom
CAP: 3 to 12 cm in diameter, convex to nearly flat, white, sometimes with a few scales, margin slightly overhanging the edge of the cap, flesh not yellowing when bruised but discoloring pinkish in age, odor and taste mild. **GILLS:** Free, pink when young, becoming dark chocolate brown. **STALK:** 2 to 7 cm long, 1 to 2 cm wide, white, tapered toward base, smooth or slightly fibrous, partial veil leaving a thin membranous ring that may disappear. **SPORES:** Dark chocolate brown, elliptical, smooth. **HABITAT:** Scattered or in arcs and rings in lawns and grassy fields, sometimes in large numbers, widespread. **EDIBILITY:** Edible and choice.

Agaricus campestris

The Meadow Mushroom is a common lawn mushroom throughout North America. Defining characteristics include the white cap, pink gills in young specimens, weakly developed ring, chocolate brown spores and gills (when mature), and occurrence in grassy areas. The color of the gills is important since the toxic **A. californicus** (p. 223) has whitish gills when young and typically has a relatively long stalk with a membranous, persistent ring. It also has a faint odor of phenol, unlike *A. campestris*, which has a mild odor. **Agaricus xanthodermus** (p. 230) also fruits in grassy areas, but it stains a distinct yellow upon bruising and has a well-developed ring. It has an odor of phenol.

Agaricus cupreobrunneus

CAP: 3 to 10 cm in diameter, convex to nearly flat, brown, covered with brown to reddish brown flattened fibrils, flesh not yellowing when bruised but discoloring pinkish in age, odor and taste mild. **GILLS:** Free, pink when young becoming dark chocolate brown. **STALK:** 2 to 7 cm long, 1 to 2 cm wide, equal or tapered toward base, white, smooth or fibrillose, partial veil leaving a thin, poorly developed ring that sometimes disappears in age. **SPORES:** Dark chocolate brown, elliptical, smooth. **HABITAT:** Scattered or in rings and arcs in grassy areas and disturbed or

Agaricus cupreobrunneus

hard-packed soil such as vacant lots and poorly maintained pastures. **EDIBILITY:** Edible.

The brown scaly cap, pink gills when young, mild odor, and penchant for growing in poor soils characterize this good edible. Like **A. campestris** (p. 224), it has a slight ring that may disappear; *A. campestris*, however, has a white or nearly white, smooth to almost smooth cap. **Agaricus augustus** (p. 221) has yellowish brown or brown fibrils on the cap, but it ages and stains yellow, eventually developing a golden hue. It has a fragrant odor of almond paste. The woodland **A. hondensis** (below) has pale pinkish brown flattened fibrils on the cap and an unpleasant odor.

Agaricus hondensis

CAP: 7 to 15 cm in diameter, convex to plane in age, white with pale pinkish brown appressed fibrils, especially in the center, the fibrils sometimes becoming scaly and pronounced in dry weather, flesh unchanging or staining pale yellow when bruised, odor of phenol (unpleasant, resinlike). **GILLS:** Free, pale pink when young and then brown and finally dark chocolate to blackish brown. **STALK:** 7 to 15 cm long, 1 to 3 cm wide, white but aging brownish, the base distinctly enlarged or even bulbous, smooth above and below the persistent, thick, felty skirt-like ring; when cut, the base has an odor of phenol and usually stains yellow. **SPORES:** Chocolate brown, elliptical, smooth. **HABITAT:** Single, in groups, or in rings in woods under both conifers and hardwoods. **EDIBILITY:** Toxic.

Agaricus hondensis is characterized by a phenol-like odor that is particularly evident when the base of the stalk is crushed, the smooth stalk both above and below the thick ring, and the pale pinkish brown flattened fibrils on the cap. **Agaricus moelleri** (*A. praeclaresquamosus*), a borrowed European name that is likely incorrect for local collections, has grayish brown appressed fibrils or scales on a cap that is often flattened in age (p. 424). It slowly develops vinaceous stains when bruised and grows in disturbed ground near paths and roads in the woods. Like *A. hondensis*, it causes gastrointestinal upsets if eaten. **Agaricus subrutilescens** (p. 229) is characterized by a purplish brown fibrillose cap, cottony scales of the lower stalk, and mild odor.

Agaricus hondensis

Agaricus silvicola

CAP: 6 to 15 cm in diameter, round becoming convex and finally plane, white, smooth or slightly fibrillose, dry, yellowing when bruised or in age, odor sweet (like almonds). **GILLS:** Free, whitish becoming pinkish gray and finally dark brown. **STALK:** 5 to 15 cm long, 1 to 3 cm wide, white but often with a yellow tinge in age, equal or enlarged at base, smooth or slightly scaly, partial veil white, membranous, leaving a large skirtlike ring high on the stalk; the underside of the ring splitting and forming patches of tissue in a spokelike pattern. **SPORES:** Chocolate brown, elliptical, smooth. **HABITAT:** Single or scattered in groups in mixed hardwood/coniferous forests. **EDIBILITY:** Edible.

Like **Agaricus hondensis** (p. 226) and **A. subrutilescens** (p. 229), this is a mushroom of woods. The white cap that slowly yellows when bruised or in age and the anise or almond odor distinguishes *A. silvicola* from the other two. The yellowing reaction is similar to that of **A. xanthodermus** (p. 230), which causes gastrointestinal distress if eaten. Therefore, caution is warranted when collecting *A. silvicola* for the table. **Agaricus semotus** is a small species (cap up to 6 cm broad) with an anise odor. It has reddish fibrils on the cap, which bruises yellow, and is not uncommon under oaks.

Agaricus silvicola

Agaricus subrutilescens

Agaricus subrutilescens
Wine-colored Agaricus

CAP: 5 to 15 cm in diameter, convex becoming plane, the surface covered with purplish brown or vinaceous appressed fibrils on a whitish background, usually darkest in the center, dry, not staining when bruised, odor and taste mild. **GILLS:** Free, white becoming pink and then finally chocolate brown. **STALK:** 5 to 15 cm long, 1 to 3 cm wide, white, equal or enlarged at base, smooth above the white skirtlike ring, covered with cottony white scales below the ring. **SPORES:** Chocolate brown, elliptical, smooth. **HABITAT:** Single or scattered in hardwood/coniferous forests. **EDIBILITY:** Edible.

Most specimens of *A. subrutilescens* are easily identified by the purplish brown, fibrillose cap, cottony scales of the stalk below the clearly defined ring, and occurrence in forests. ***Agaricus hondensis*** (p. 226) is similar in size and also occurs in the woods. It can be identified by the pinkish brown appressed fibrils of the cap, the odor of phenol in the base of an enlarged or bulbous stalk, and the smooth stalk surface both above and below the thick feltlike ring. The base of the stalk of *A. hondensis* may stain yellow upon bruising. Unlike *A. subrutilescens*, it is toxic. ***Agaricus moelleri*** (*A. praeclaresquamosus*) (p. 424) has gray to grayish brown fibrils or scales a flat cap and develops vinaceous stains when bruised.

Agaricus xanthodermus

Agaricus xanthodermus

CAP: 5 to 15 cm in diameter, marshmallow shaped to convex to nearly plane, margin inrolled at first, white or white with a tan center, dry, usually smooth, staining bright yellow when bruised and then becoming brownish as the yellow fades (the yellow color will develop quickly if the margin of the cap is rubbed with a finger), odor of phenol (unpleasant, resinlike). **GILLS:** Free, whitish when young, becoming pink and then chocolate brown. **STALK:** 5 to 12 cm long, 1 to 3 cm wide, equal or with an enlarged base, smooth, bruising bright yellow at the base, the partial veil leaving a thick, white membranous ring that has cottony patches on the lower surface. **SPORES:** Chocolate brown, elliptical, smooth. **HABITAT:** Scattered to gregarious in lawns, under trees, and along roads. Widely distributed and common. **EDIBILITY:** Poisonous, causing gastrointestinal upsets.

The bright yellow color that quickly develops when any part of this mushroom is bruised is distinctive. The phenolic odor, which is especially pronounced at the base of the stalk, the persistent membranous ring on the stalk, and white gills in young specimens are also key characteristics. ***Agaricus californicus*** (p. 223) is similar, but it does not change colors or will develop a faint yellow color upon bruising. *Agaricus* species with a phenolic odor should not be collected for the table.

Crepidotus mollis

CAP: 1 to 5 cm broad, fan or kidney shaped, convex to plane, gelatinous when wet, cream, buff, to yellow-brown with brown fibrils and scales on the cap surface (but the fibrils and scales wear away and thus the cap appears smooth), odor and taste mild. **GILLS:** Whitish or buff becoming brown as the spores mature, close, radiating from the point of attachment. **STALK:** Absent or just a point of attachment to substrate. **SPORES:** Brown, elliptical, smooth. **HABITAT:** Single, scattered, or in groups on dead wood or the bark of living trees, usually hardwoods. **EDIBILITY:** Unknown.

The flabby texture of the fan-shaped fruiting body, hairy cap, and brown spores are the important field marks. It grows singly or in overlapping shelves like the Oyster Mushroom, ***Pleurotus ostreatus*** (p. 132), but that mushroom is larger and has a smooth cap surface and white, not brown, spores. Similar species include ***Crepidotus applanatus,*** which has a whitish, velvety to smooth, hygrophanous cap that becomes brown as it dries or in age, and brown spores that are minutely roughened.

Crepidotus mollis

It grows on dead hardwood logs and stumps. ***Crepidotus crocophyllus*** has reddish brown fibrillose scales on the surface of the cap and yellow to orange tones on the gills before they become brown with spores, which are also minutely roughened.

Paxillus involutus

CAP: 5 to 15 cm in diameter, convex to plane with an inrolled margin and a depressed center, striate, yellow-brown, brown, or reddish brown, darker brown where bruised, the surface minutely fibrillose but smooth in age, viscid in moist weather, flesh yellow-brown, bruising brown, odor mild, taste sour. **GILLS:** Decurrent, close, yellow to light brown, staining dark brown, often forked or veined near the stalk. **STALK:** 3 to 6 cm long, 1 to 3 cm wide, equal or tapering toward the base, smooth, central or eccentric, colored like cap or paler, bruising brown, partial veil absent. **SPORES:** Yellow-brown, elliptical, smooth. **HABITAT:** Scattered or gregarious on ground in hardwood/coniferous forests and under ornamental birch. **EDIBILITY:** Toxic.

The brown cap with a strongly inrolled margin (conspicuous in all but mature specimens), decurrent gills that are forked or veined near the stalk, and brown stains are distinctive field char-

Paxillus involutus

acteristics. The stalk is usually shorter than the cap is wide. **Tapinella atrotomentosa** (*Paxillus atrotomentosus*) is similar, but it has a velvety cap and stalk and fruits on decaying wood. **Tapinella** (*Paxillus*) **panuoides** is fan shaped and stalkless or nearly so and grows laterally from decaying coniferous wood. Like *P. involutus*, its gills are forked or veined near the base of the cap. **Phylloporus arenicola**, a gilled bolete, has a dry brown cap, decurrent bright yellow gills, and yellow-brown spores. It grows on the ground.

Psathyrella candolleana

CAP: 2 to 7 cm in diameter, broadly convex, sometimes umbonate, honey colored or tan and fading in age, smooth, hygrophanous (fading as it loses moisture), the margin edged with white partial veil fragments at least when young, fragile, odor and taste mild. **GILLS:** Adnate, close, whitish and then grayish purple and finally brown as the spores mature. **STALK:** 4 to 10 cm long, 3 to 7 mm wide, equal, white, fragile, hollow, partial veil leaving remnants on cap margin, otherwise disappearing and almost never leaving a ring on the stalk. **SPORES:** Dark brown, elliptical, smooth. **HABITAT:** Scattered, gregarious, or in small clusters in landscaped areas, lawns, gardens, and on buried wood, in urban

Psathyrella candolleana

areas, and along roads and paths in woods. **EDIBILITY:** Edible but of poor quality and easily misidentified.

This common urban mushroom is identified by its honey-colored cap, white fragile stalk, and dark brown spores. It sometimes fruits in large numbers. ***Psathyrella gracilis*** (below) is also common in landscaped areas. It differs from *P. candolleana* by a browner, smaller, and more conical cap (which is hygrophanous and fades in age), faintly striate cap margin, and the absence of a partial veil (and thus any remnants on the margin like *P. candolleana*). It also fruits in large numbers on wood chips and other woody debris in moist weather.

Psathyrella gracilis

CAP: 1 to 4 cm in diameter, broadly conical to bell shaped when young, becoming convex to flat at maturity, the surface moist when young but soon dry, brown at first, hygrophanous, becoming buff to tan in age as it dries, the margin often, but not always, striate, fragile, odor and taste mild. **GILLS:** Adnexed to adnate, close, buff becoming grayish brown to brown as the spores mature. **STALK:** 4 to 12 cm long, 1 to 3 mm wide, equal, straight, fragile, whitish to buff, partial veil absent. **SPORES:** Purple-brown, elliptical, smooth. **HABITAT:** Densely gregarious or in

Psathyrella gracilis

small clusters on wood chips, woody debris, leaf litter, and disturbed soil in landscaped areas, lawns, and gardens; especially common in urban areas; sometimes fruiting in large numbers. **EDIBILITY:** Unknown but insubstantial in any case.

This is a common mushroom of wood chips and woody debris in urban parks, camping sites, and landscaped areas. The principal diagnostic features of *P. gracilis* include the hygrophanous nature of the broadly conical or bell-shaped brown cap, penchant for fruiting in large numbers, thin fragile stalk, and purple-brown spores. It sometimes fruits with **Tubaria furfuracea** (p. 251). See **Psathyrella candolleana** (p. 233) for another common species in woody debris in urban areas.

Psathyrella piluliformis

(Other name: *Psathyrella hydrophila*)

CAP: 2 to 5 cm in diameter, broadly convex, chestnut to reddish brown, smooth, hygrophanous, fading in age and in dry weather to tan or buff, the margin edged with white veil fragments. **GILLS:** Adnate, crowded, tan becoming dark brown as the spores mature. **STALK:** 2 to 7 cm long, 3 to 7 mm wide, equal or enlarged slightly at the base, white, smooth, hollow, fragile, partial veil leaving remnants on the cap margin and occasionally a faint

Psathyrella piluliformis

zone of hairs on the stalk. **SPORES:** Dark brown, elliptical, smooth. **HABITAT:** In large clusters on hardwood stumps and buried wood. **EDIBILITY:** Unknown.

The large clusters at the base of hardwood stumps, brown spores, smooth reddish brown cap, and fragile white stalk are the principal field marks. The cap margin is edged with white veil fragments that become brown from falling spores. Other woodland *Psathyrella* species and look-alikes include *P. longipes,* a leaf litter inhabitant with a broadly conical, hygrophanous brown cap that conspicuously fades from the center to the margin, which is edged with veil fragments; *P. longistriata,* with a striate margin on the brown cap and a persistent striate ring on the stalk; *Parasola conopilus* (*Psathyrella conopilus*), (pp. 64–65) commonly encountered growing gregariously in parks, grassy areas, and along paths in the woods, which has reddish brown caps with microscopic stiff hairlike cells and long thin stalks; and *Psathyrella carbonicola,* one of many mushrooms that grow on charred soil and wood ash, which has a brown cap covered at first with white fibrils.

Pholiota spumosa

CAP: 2 to 8 cm in diameter, convex to nearly flat to slightly umbonate, yellow-brown or tan with a yellowish margin, smooth, viscid when moist, flesh yellowish, odor mild. **GILLS:** Adnate or notched, close, yellow becoming brown. **STALK:** 3 to 8 cm long, 4 to 8 mm wide, more-or-less equal, fibrillose, yellow, brownish orange, or brown, partial veil pale yellow, leaving scant tissue on the cap margin and stalk. **SPORES:** Brown to reddish brown, elliptical, smooth. **HABITAT:** Scattered, gregarious, or in small clusters on decaying coniferous wood, including wood chips in landscaped areas. **EDIBILITY:** Unknown.

Other *Pholiota* species with smooth caps (or with just a few scales) include *P. astragalina,* which has a smooth, bright pinkish orange viscid cap up to 5 cm broad, yellow gills, and smooth stalk and grows on rotting conifers in the woods; *P. velaglutinosa,* with a viscid reddish brown cap, a glutinous ring on the stalk, and small scales on the stalk below the ring, which grows singly or scattered under pines; *P. highlandensis,* which has a small, smooth, brown viscid cap and fibrillose stalk and grows in dense groups on burnt wood or soil; and *P. alnicola,* with a

Pholiota spumosa

smooth, lubricious, yellow-orange cap spotted brown in age and a fibrillose stalk, which grows in clusters on exposed or buried wood in forests.

Pholiota squarrosa

CAP: 3 to 12 cm in diameter, convex to bell shaped, pale yellow to buff with darker brown erect scales, the surface dry, the margin often edged with veil remnants, at least when young, flesh white to pale yellow, odor mild or garliclike. **GILLS:** Adnexed to slightly decurrent, close, whitish to greenish yellow, becoming brown. **STALK:** 4 to 12 cm long, up to 1.5 cm wide, equal, dry, colored like cap, covered with brown scales below the ring, partial veil leaving a fragile ring. **SPORES:** Brown, elliptical, smooth. **HABITAT:** In clusters on wood in the mountains, commonly at the base of trees. **EDIBILITY:** Edible to some but disagreeable to others; not recommended.

Pholiota squarrosa grows in clusters on living and dead hardwood and conifer trees. It has erect scales both on the cap and stalk, greenish tinged gills, and a mild or garliclike odor. Unlike most species of *Pholiota*, the cap is not viscid. **Pholiota terrestris** grows in clusters on the ground. Like many other species of *Pholiota*, it has brown spores, the cap is some shade of brown,

Pholiota squarrosa

and it has scales on the stalk below the ring. ***Pholiota aurivella*** has a bright, golden orange viscid cap with spotlike brown scales and a scaly stalk below the ring. It grows in large clusters on live trees and downed logs. ***Pholiota populnea*** (*P. destruens*) has a large white or cream-colored cap and a white scaly stalk. It grows on cottonwood and aspen logs.

Agrocybe pediades
(Other name: *Agrocybe semiorbicularis*)
CAP: 1 to 3 cm in diameter, rounded to convex to nearly flat in age, pale yellow- or orange-brown becoming tan, smooth but often with superficial cracks, tacky in moist weather but soon dry, the margin sometimes edged with white veil remnants, odor and taste mild or mealy. **GILLS:** Adnate to adnexed, cream colored becoming brown. **STALK:** 2 to 5 cm long, 1 to 3 mm wide, equal, smooth, dry, cream colored, partial veil quickly disappearing, leaving traces on the edge of the cap but none on the stalk. **SPORES:** Brown, elliptical, smooth. **HABITAT:** Scattered or gregarious in lawns, pastures, and other grassy areas, and on dung. **EDIBILITY:** Not recommended.

Agrocybe pediades

This common lawn mushroom is identified by its yellow-brown to buff-colored cap, small size, brown spores, and preference for grassy areas, where it sometimes fruits in great numbers. The cap of at least some specimens in a collection will be superficially cracked. Related species include **A. praecox**, which has a larger cream-colored or yellow-brown cap (3 to 10 cm in diameter) and a partial veil that leaves remnants of the cap margin, with or without an ephemeral ring on the stalk. The cap surface is also superficially cracked in age. It often grows in spring or early summer on disturbed ground, along roads, and in wood chips. **Agrocybe putaminum** has a dull brownish orange cap, lacks a veil, and is common on wood chips.

Hebeloma crustuliniforme
Poison Pie

CAP: 3 to 8 cm in diameter, convex becoming plane, buff (the color of pie crust), viscid when moist, smooth, odor radishlike. **GILLS:** Adnate or notched, close, minutely serrated (like the blade of a saw), beaded with droplets of water in wet weather, tan to brown. **STALK:** 4 to 9 cm long, 0.5 to 1.5 cm wide, equal with a wider base, colored like cap or paler, solid, the apex pruinose (covered with whitish powdery granules), partial veil ab-

Hebeloma crustuliniforme

sent. **SPORES:** Brown, elliptical, slightly roughened. **HABITAT:** Single, scattered, or in arcs and rings on ground in hardwood and coniferous forests or under trees and shrubs (e.g., manzanitas) in landscaped areas. **EDIBILITY:** Poisonous, causing gastrointestinal upsets.

The field marks of this common mycorrhizal species are the piecrust color of the smooth cap, brown spores, powdery granules on the stalk just beneath the cap, and absence of a ring. Relatives include *H. sinapizans*, a more robust mushroom with a reddish brown cap 15 cm or more broad (often with a wavy margin) and scaly stalk up to 3 cm wide, and *H. insigne*, which has a brown to pinkish tan cap, brown gills, and a scaly bulbous stalk. Both of these grow under hardwoods and conifers; the latter is especially common under aspen.

Inocybe Species

These are small to medium-sized, brown-spored mushrooms that grow on the ground in mycorrhizal association with many trees. The caps are usually conical when young and retain an umbo even if they broadly flatten in age. The surface of the cap is dry and fibrillose, silky, or scaly. The fibrils and scales, if present, are radially arranged, resulting in characteristic radial cracks along the cap margin as the mushroom matures. Red or orange stains and distinctive odors aid in the identification of some species. Because the toxin muscarine is often present in dangerous amounts, no species of *Inocybe* should be considered for the table.

a. Cap and stalk staining reddish
   b. Cap nearly smooth, white bruising pink or orange; odor unpleasant..**Inocybe pudica**
   b. Cap surface radially fibrillose
      c. Cap surface buff with brown fibrils; odor mild to mealy.
      ....**I. adaequata** (see *I. pudica*)
      c. Cap surface yellow-brown; odor fruity ....
      ....**I. fraudans** (see *I. pudica*)

a. Cap and stalk not staining red
   d. Cap white..**I. geophylla**
   d. Cap lilac ....**I. lilacina** (see *I. geophylla*)
   d. Cap yellow to brownish
      e. Cap straw colored, fibrillose but not scaly; odor of green corn ....**I. sororia**
      e. Cap brown and covered with hairy scales; odor fruity ....
      ....**I. flocculosa** (see *I. sororia*)

Inocybe flocculosa

Inocybe geophylla

CAP: 2 to 4 cm in diameter, conical to bell shaped, becoming umbonate in age, splitting along the margin, white, silky to smooth, dry, odor unpleasant. **GILLS:** Adnexed to notched, close, whitish becoming brown as the spores mature. **STALK:** 2 to 6 cm long, 2 to 5 mm wide, equal, white, dry, silky, partial veil fibrillose, leaving a zone of hairs on the stalk or disappearing. **SPORES:** Dull brown, elliptical, smooth. **HABITAT:** Scattered or gregarious on ground in mycorrhizal association with hardwoods and conifers. **EDIBILITY:** Poisonous.

The overall white color, small size, conical or umbonate cap, and brown spores help define this common mushroom of the forest. *Inocybe lilacina* is similar in all respects but is lilac in color. *Inocybe pudica* (p. 243) is also white like *I. geophylla*, but it stains reddish orange in age. The brown spores quickly distinguish these small species of *Inocybe* from pale-colored or lilac-colored species of *Mycena*, which have conical striate caps, thin stalks, and white spores. The lilac color of *I. lilacina* could be confused with **Clitocybe nuda** (p. 151) or many lilac species of *Cortinarius*, but those species are much larger and have pink or rusty brown spores, respectively.

Inocybe geophylla

Inocybe pudica

Inocybe pudica

CAP: 2 to 4 cm in diameter, conical becoming convex to plane with an umbo, smooth at first, later silky, white staining pink or orange in age, flesh pale pink when exposed, odor unpleasant. **GILLS:** Adnexed to notched, whitish with pink or orange stains, becoming brown as the spores mature. **STALK:** 2 to 6 cm long, 0.5 to 1 cm wide, equal or wider at the base, white with pink or orange stains, pruinose at the apex, otherwise mostly smooth, partial veil fibrillose and white, disappearing or leaving a zone of hairs. **SPORES:** Dull brown, elliptical, smooth. **HABITAT:** Scattered on the ground under conifers. **EDIBILITY:** Poisonous, containing muscarine.

The pink to orange stains on the white fruiting body and dull brown spores are a helpful set of features. Other *Inocybe* species that develop reddish colors include the oak-loving *I. adaequata* (*I. jurana*), which has a buff, vinaceous-tinged cap with brown fibrils and a mild to mealy odor, and *I. fraudans* (*I. pyriodora*), with a yellowish brown fibrillose cap and stalk that develop reddish brown stains where injured and a fruity odor. The brown spores of *Inocybe* separate these species from the more robust reddish streaked, white-spored **Hygrophorus russula** (p. 130).

Inocybe sororia

CAP: 2 to 7 cm in diameter, conical or bell shaped becoming flattened in age but retaining an umbo, yellow to brownish yellow, dry, distinctly radially fibrillose, the margin often splitting in age, odor of green corn. **GILLS:** Adnate to adnexed, close, yellow becoming brown, the edges lighter in color. **STALK:** 3 to 10 cm long, 4 to 8 mm wide, equal or slightly enlarged at the base, white to yellow, dry, fibrillose, partial veil absent. **SPORES:** Dull brown, elliptical, smooth. **HABITAT:** Single or scattered on ground in hardwood/coniferous forests. **EDIBILITY:** Poisonous.

Inocybe sororia is recognized by its yellowish (straw-colored), fibrillose, conical or umbonate cap that often splits along the margin some distance toward the center, brown spores, and odor of green corn. **Inocybe adaequata** (*I. jurana*) is very similar in size and shape but develops reddish or vinaceous colors

Inocybe sororia

on the cap and stalk and has a mild to mealy odor. It favors oaks. The surface of the cap of *I. brunnescens* is brown and has a white persistent veil patch. The stalk has an abrupt white bulb at the base. *Inocybe flocculosa* (p. 241) has a brown hairy cap and smooth stalk, and *I. maculata* has a brown fibrillose cap and stalk. The latter two have a fruity odor. All *Inocybe* species should be considered poisonous because most contain dangerous concentrations of the toxin muscarine.

Gymnopilus junonius

(Other names: *Gymnopilus spectabilis*) Big Laughing Jim

CAP: 8 to 30 cm or more in diameter, convex to plane, yellow-orange to rusty orange, dry, smooth or silky fibrillose, flesh yellow, odor mild, taste bitter. **GILLS:** Adnate to slightly decurrent, close, yellowish becoming rusty orange. **STALK:** 5 to 25 cm long, 1 to 5 cm wide, equal, clavate, or enlarged in the middle, often tapered at the base, colored like the cap, solid, firm, dry, streaked with fibers below the ring, partial veil leaving a yellow or orange ring that may disappear. **SPORES:** Rusty orange, elliptical, roughened. **HABITAT:** In clusters on logs, buried wood, and especially at the base of stumps of both conifers and hardwoods. **EDIBILITY:** Inedible; some forms are hallucinogenic.

Gymnopilus junonius

The large size of the orange caps, clustered growth habit, rusty orange spores, and ring on the stalk identify this species. The clusters reach impressive sizes and are eye-catching from long distances. The above description may apply to a complex of similar species in the western United States. *Gymnopilus junonius* might be confused with **Armillaria mellea** (p. 134), which also grows in dense clusters on wood, but that species has honey-colored caps and white spores; and **Omphalotus olivascens** (p. 143), which is orange and grows in clusters on wood, but lacks a veil and has white spores.

Gymnopilus sapineus

CAP: 2 to 8 cm in diameter, rounded to convex to nearly flat, yellow-orange or brownish orange, fading in age, dry, smooth or minutely scaly, sometimes developing shallow cracks in age, flesh yellow, odor mild, taste bitter. **GILLS:** Adnate, yellow or yellow-orange becoming rusty orange. **STALK:** 3 to 7 cm long, up to 1 cm wide, equal, yellowish or yellow-orange, becoming rusty or brownish orange when bruised or in age, smooth or minutely fibrillose, white mycelium often adhering to the base, partial veil leaving a few hairs near the top of the stalk or disappearing alto-

Gymnopilus sapineus

gether. **SPORES:** Rusty orange, elliptical, warted. **HABITAT:** Single, scattered, or in small clusters on coniferous wood, wood chips, and lignin-rich soil; sometimes fruiting in large numbers. **EDI-BILITY:** Unknown.

This description may represent more than one species, but as a group they can be identified by their orange color, dry cap, rusty orange spores, and growth on wood. **Gymnopilus pene-trans** may be a synonym; if not, *G. penetrans* differs from *G. sapineus* by microscopic characteristics of the cap surface. Many other species of *Gymnopilus* grow on wood, including some that grow on wood chips in landscaped areas. The latter group includes **G. aeruginosus,** which has conspicuous bluish green stains on the cap, stalk, and interior, and **G. luteofolius,** which has a vinaceous scaly cap that yellows in age.

Galerina marginata
(Other name: *Galerina autumnalis*)

CAP: 1 to 4 cm in diameter, convex to nearly flat, yellow-brown to brown becoming tan, smooth, viscid when moist, the margin finely striate when fresh, odor mealy. **GILLS:** Adnexed to slightly decurrent, pale yellow-orange becoming brown in age. **STALK:** 2

Galerina marginata

to 8 cm long, 3 to 6 mm wide, equal or wider at base, tan or brown, dry, hollow, smooth above the ring and fibrillose below the ring, partial veil leaving a fragile white ring, which may disappear in age. **SPORES:** Rusty brown, elliptical, roughened, inamyloid. **HABITAT:** Scattered, gregarious, or in small clusters on rotting wood. **EDIBILITY:** Deadly poisonous.

Fruiting bodies of *G. marginata* contain amanitins, the deadly toxins found in **Amanita phalloides**, the Death Cap (p. 77). *Galerina marginata* is identified by its small size, brown cap, growth on wood, ringed stalk, and rusty brown spores. The margin of the smooth cap may be striate, at least when the mushroom is moist. Many other small or smaller species of *Galerina* are very difficult to identify in the field. Many have yellow-brown or orange caps and partial veils that do not leave a ring on the stalk. All have rusty brown spores. Common habitats of many *Galerina* species are moss and well-rotted wood.

Bolbitius titubans

(Other name: *Bolbitius vitellinus*) Sunny Side Up

CAP: 2 to 7 cm in diameter, oval when young becoming conical, convex, or sometimes plane in age, bright yellow but soon paler in age and becoming gray or white, viscid, the margin striate, very fragile. **GILLS:** Adnate to adnexed, close, white to yellow becoming light brownish orange as the spores mature. **STALK:** 5 to 12 cm long, 2 to 6 mm wide, equal, extremely fragile, white to pale yellow, partial veil absent. **SPORES:** Rusty brown, elliptical, smooth. **HABITAT:** Single or scattered in lawns, dung, and disturbed areas during moist weather. **EDIBILITY:** Unknown but too small to be of value.

The bright yellow viscid cap, rusty brown spores, and extremely fragile texture are the principal field marks of *B. titubans.* Specimens often wither away within a single day. In fact, the bright yellow color of the cap often needs to be seen in the morning. **Bolbitius aleuriatus** is an attractive woodland relative with a small, fragile viscid cap (less than 3 cm in diameter). When fresh, the deeply striate cap is lilac and the stalk is yellow. It occurs on rotting wood.

Bolbitius titubans

Conocybe apala
(Other names: *Conocybe albipes, C. lactea*) Dunce Cap
CAP: 1 to 2.5 cm in diameter and in height, conical, sometimes with an uplifted margin, cream colored or buff, dry, striate and often wrinkled in age, very thin and fragile, odor and taste mild. **GILLS:** Adnexed or almost free, close, whitish becoming brownish orange. **STALK:** 3 to 10 cm long, 1 to 2 mm wide, equal, white, very fragile, hollow, partial veil absent. **SPORES:** Cinnamon brown, elliptical, smooth. **HABITAT:** Scattered to gregarious in lawns and other grassy areas, often in summer and fall. **EDIBILITY:** Unknown but too small to be considered in any case; some *Conocybe* species are poisonous (see below).

This common lawn mushroom is characterized by its cream-colored conical cap, long, fragile stalk, and cinnamon brown spores. Warm humid weather and a good irrigation will sometimes result in large fruitings in turf. *Conocybe tenera* is similar, but it has a brown hygrophanous cap with a striate or smooth

Conocybe apala

surface and a brown stalk. It also grows in lawns but is not restricted to that habitat. **Conocybe filaris** (*Pholiotina filaris*), which grows in grassy areas and on woody debris, has a brown cap up to about 2 cm broad and a membranous striate ring on the stalk. It is noteworthy because it contains amanitins, the deadly toxins present in **Amanita phalloides**, the Death Cap (p. 77).

Phaeocollybia californica

CAP: 2 to 8 cm in diameter, conical, retaining a distinct umbo in age, bright orange to orange-brown becoming reddish brown in age, smooth, viscid, hygrophanous (fading as it loses moisture), odor mild to pungent. **GILLS:** Adnate, adnexed, or free, pale orange becoming rusty brown as the spores mature. **STALK:** 6 to 15 cm long (plus an additional 10 to 20 cm below ground), 0.5 to 1 cm wide, equal above, gradually tapering toward the base, dark orange to reddish brown, often paler toward the apex, smooth, cartilaginous, hollow, partial veil absent. **SPORES:** Cinnamon brown, lemon shaped, ornamented with minute warts. **HABITAT:** Densely gregarious on ground under conifers, often in large numbers. **EDIBILITY:** Unknown.

The cone-shaped cap, long rooting stalk, cinnamon brown spores, and gregarious growth habit make this species easy to identify. **Phaeocollybia kauffmanii** is larger (cap to 15 cm or

Phaeocollybia californica

more broad) but otherwise similar. The stalk is 3 cm or more wide and densely stuffed with whitish mycelium. ***Phaeocollybia olivacea*** has viscid, olive green conical caps and olive stalks tinged reddish near ground level and below the soil surface. The stalk in all of these species is deeply rooted and appears like a taproot. ***Caulorhiza umbonata*** (p. 173) also has a long taproot but has white spores.

Tubaria furfuracea

CAP: 1 to 4 cm in diameter, convex to plane to centrally depressed, reddish brown or orange-brown fading to tawny-brown or paler, dotted on the margin with small white patches of veil tissue (but they often disappear in age or in rainy weather), hygrophanous (fading as it loses moisture), the margin faintly striate, odor mild. **GILLS:** Adnate to slightly decurrent, brown. **STALK:** 1 to 5 cm long, 2 to 4 mm wide, equal, brown or tan, hollow, partial veil leaving patches on cap margin, otherwise disappearing. **SPORES:** Pale rusty brown, elliptical, smooth. **HABITAT:** Scattered or gregarious on sticks, wood chips, thatch in grassy areas, and woody debris. **EDIBILITY:** Unknown.

When young, this modestly sized mushroom can be recognized by the white pieces of veil tissue on the reddish brown

Tubaria furfuracea

caps. It often fruits in large numbers on woody debris like wood chips but is frequently encountered growing in grassy areas as well. **Tubaria confragosa** is similar but slightly larger, and it has an ephemeral ring on the stalk. It grows in clusters on woody debris in forests. Several other smaller, reddish to brown *Tubaria* species occur on woody debris or disturbed soils.

Cortinarius Species

In terms of number of species, this is the largest genus of gilled mushrooms in the western states. They are united by rusty brown spores and a cobweblike partial veil called a cortina. Very often the cortina remains as a faint band of fibers around the stalk. The stalk is equal, tapered, or strongly bulbous. The genus is often loosely divided into several subgenera based on the color and texture of the cap and stalk, presence of water-soluble pigments, and spore characteristics. *Cortinarius* species are mycorrhizal with conifers, birches, oaks, and other trees. Because some species contain dangerous amounts of the toxin orellanine and specimens are notoriously difficult to identify to species, none should be considered for the table.

a. Cap viscid, at least when moist
  b. Stalk viscid
    c. Stalk very wide (4 to 7 cm wide); cap yellow-brown .... **Cortinarius ponderosus**
    c. Stalk mostly less than 3 cm wide
      d. Stalk uniformly white; cap orange; taste of cap bitter ... . **C. vibratilis**
      d. Stalk not all white
        e. Stalk with brownish bands, cap yellow-brown ....**C. trivialis**
        e. Stalk whitish with purple slime; cap dark brown ....**C. vanduzerensis**
  b. Stalk dry
    f. Stalk with a basal bulb
      g. Cap with lilac or olive tones; gills bluish when young ....**C. glaucopus**
      g. Cap yellowish; gills yellow when young........................ .... **C. xanthodryophilus**
    f. Stalk without a basal bulb; odor fruity
      h. Cap yellowish...**C. percomis**
      h. Cap lavender ....**C. subfoetidus**

a. Cap dry
  i. Partial veil persistently attached to cap margin.................... ....**C. verrucisporus**
  i. Partial veil not persistently attached, leaving hairs on the stalk
    j. Cap covered with tufts of hair
      k. Cap deep purple ....**C. violaceus**
      k. Cap yellow-brown.......................................**C. cotoneus**
    j. Cap smooth or silky
      l. Cap silvery lilac or lilac
        m. Cap and stalk silvery lilac; flesh whitish or pale lilac .... **C. griseoviolaceus**
        m. Cap and stalk lilac; flesh mottled reddish brown .... **C. traganus**
      l. Cap not predominately lilac but may have lilac tones
        n. Cap yellow-brown; immature gills brightly colored ....**C. croceus**
        n. Cap predominately brown; immature gills not brightly colored

o. Cap purplish brown becoming brown; stalk whitish tinged violet.............................**C. evernius**

o. Cap orange-brown; stalk pale brown with a swollen base...**C. laniger**

Cortinarius cotoneus

CAP: 4 to 10 cm in diameter, rounded when young, soon convex to plane, olive brown to yellow-brown, darker in the center, the surface feltlike from a covering of dark short fibrils, dry, flesh buff to brownish, sometimes tinged olive, odor radishlike. **GILLS:** Adnate to notched, olive when young becoming brownish and finally cinnamon brown as the spores mature. **STALK:** 5 to 10 cm long, 1 to 2 cm wide, equal or wider at the base, dry, colored like cap or paler toward the apex, partial veil leaving a zone of hairs on the upper stalk. **SPORES:** Rusty brown, globose or nearly so, roughened with warts. **HABITAT:** Single, scattered, or gregarious on ground in hardwood/coniferous forests. **EDIBILITY:** Unknown but not recommended since some species of *Cortinarius* are deadly poisonous.

In general, *Cortinarius* species in the subgenus *Leprocybe* have olive yellow or olive brown colors, dry caps, and a radish-like odor. They also fluoresce a bright yellow in ultraviolet light,

Cortinarius cotoneus

hardly a field characteristic but an interesting feature of this group. *Cortinarius cotoneus* is quite distinct by the feltlike cover on the cap. **Cortinarius subalpinus** is commonly encountered in spring in the Sierra Nevada. It has an overall yellow-brown color and a persistent ring. **Cortinarius rubicundulus** has a finely fibrillose whitish to buff cap (3 to 9 cm broad) and stalk partially covered with reddish veil tissue. The flesh stains yellow, resulting in a distinctive color combination.

Cortinarius croceus

CAP: 2 to 8 cm in diameter, convex becoming plane or broadly umbonate, yellow-brown or reddish brown, the margin often yellow, dry, silky to fibrillose, flesh yellow, odor mild or radish-like. **GILLS:** Adnexed to notched, yellow or tinted orange, becoming rusty as the spores mature. **STALK:** 3 to 8 cm long, up to 1 cm wide, equal, dry, silky or fibrillose, yellowish, partial veil leaving a few hairs on the upper stalk but usually disappearing. **SPORES:** Rusty brown, elliptical, minutely roughened. **HABITAT:** Single, scattered, or gregarious on ground in hardwood/coniferous forests. **EDIBILITY:** Unknown but not recommended since some species of *Cortinarius* are deadly poisonous.

Cortinarius croceus

This species represents the subgenus *Dermocybe*, a collection of lively colored *Cortinarius* species. These mushrooms are well known for their bright pigments that are soluble in alcohol and water; thus, they are highly sought after for dyeing fabrics. Similar species include **C. phoeniceus var. occidentalis,** which has a maroon-colored cap, red gills, and a yellow stalk; **C. cinnamomeus** (p. 422), with a yellow-brown or olive yellow cap that develops cinnamon colors in age, and yellow-orange gills; and **C. californicus**, a similar but unrelated species with a rusty red cap and stalk and bright rusty orange gills.

Cortinarius evernius

CAP: 3 to 9 cm in diameter, bell shaped when young becoming broadly umbonate, reddish to purplish brown fading as it dries (hygrophanous), the margin often edged white, smooth or appearing silky, dry, flesh violet but fading in age, odor mild. **GILLS:** Adnate or notched, violet at first, becoming rusty brown. **STALK:** 7 to 15 cm long, 1 to 2 cm wide, equal to clavate, dry, tinged violet, covered in part with whitish veil remnants, partial veil leaving a few hairs on the stalk but often disappearing altogether. **SPORES:** Rusty brown, elliptical, minutely warted. **HABITAT:** Sin-

Cortinarius evernius

gle, scattered, or in groups on ground under conifers. **EDIBILITY:** Unknown but not recommended since some species of *Cortinarius* are deadly poisonous.

This violet-brown *Cortinarius* is a representative of the large subgenus *Telamonia.* As a group they are generally defined by a dry, often brown, hygrophanous cap and dry stalk. Other species in the subgenus *Telamonia* include **C. brunneus** (in the broad sense, i.e., representing a group), with a smooth dark brown cap that fades to light brown as it dries, cap margin edged in white, and pale brown stalk that may be whitish from remnants of the veil; **C. obtusus,** which has an umbonate, hygrophanous brown cap 1 to 4 cm broad and a fragile white stalk and is common under conifers; and **C. vernus,** an example of several species with a dull brown, slightly umbonate cap and a pale-colored stalk.

Cortinarius glaucopus

CAP: 4 to 12 cm in diameter, convex, color variable, bluish to greenish gray, in age becoming yellow-brown to rusty brown with a paler margin, the surface covered with yellowish or darker fibrils, causing a streaked or silky appearance, viscid when moist, flesh firm, buff or tinged blue, odor mild. **GILLS:** Adnate or notched, blue or violet when young becoming brown or rusty brown as the spores mature. **STALK:** 4 to 10 cm long, 1 to 3 cm wide and equal above the rimmed basal bulb, the surface of the stalk blue or violet but paler toward the base, aging brown, dry, firm, the partial veil leaving a zone of hairs on the stalk that becomes brown as it catches falling spores. **SPORES:** Rusty brown, elliptical, minutely roughened by small warts. **HABITAT:** Scattered or gregarious on the ground in hardwood/coniferous forests. **EDIBILITY:** Unknown but not recommended since some species of *Cortinarius* are deadly poisonous.

This common mycorrhizal mushroom of oaks and other trees is identified by the enlarged base of the stalk, bluish gills, and variably colored innately fibrillose cap. Examples of other bluish species with rimmed basal bulbs include **C. caerulescens,** which has a silky bluish violet cap and stalk, and **C. sodagnitus,** which has a bitter-tasting, bald cap that develops ochraceous colors. Color reactions of the cap cuticle in potassium hydroxide are used to distinguish these and their many rela-

Cortinarius glaucopus

tives—brown, yellow, and red in *C. glaucopus*, *C. caerulescens*, and *C. sodagnitus*, respectively.

Cortinarius griseoviolaceus
(Other name: *Cortinarius alboviolaceus*)
CAP: 3 to 8 cm in diameter, bell shaped to convex to umbonate, dry but sticky when moist, silky and fibrillose, pale lilac when young, silvery lilac, white, or grayish with just a hint of lilac in age, flesh whitish or pale lilac, odor and taste mild. **GILLS:** Adnate or adnexed, lilac or grayish lilac, becoming rusty brown as the spores mature. **STALK:** 4 to 8 cm long, 0.5 to 1.5 cm wide at the apex but enlarged toward the base, colored like the cap, dry, fibrillose-silky, the base sheathed with white veil tissue from the universal veil, partial veil leaving a zone of hairs or disappearing. **SPORES:** Rusty brown, elliptical, ornamented with warts. **HABITAT:** Scattered or in small clusters on the ground under conifers. **EDIBILITY:** Unknown but not recommended since some species of *Cortinarius* are deadly poisonous.

Both the cap and stalk of this mushroom are silvery lilac, aging silvery white, with a silky dry surface (characteristics of the subgenus *Sericeocybe*). **Cortinarius alboviolaceus,** which is essentially identical, occurs in eastern North America. It report-

Cortinarius griseoviolaceus

edly favors hardwoods, whereas *C. griseoviolaceus* favors conifers. These species could be confused with **Inocybe lilacina** (p. 242), but that species is smaller, has brown spores rather than rusty brown spores, and has an odor of green corn.

Cortinarius laniger

CAP: 3 to 12 cm in diameter, bell shaped becoming broadly convex or umbonate, eventually plane, pale cinnamon brown becoming dull orange-brown or tan in age, smooth, dry, hygrophanous, flesh white or light brown, odor often radishlike, taste mild. **GILLS:** Adnate, adnexed, or notched, cinnamon colored becoming rusty brown. **STALK:** 4 to 10 cm long, 1 to 3 cm wide, clavate and swollen at the base, colored like cap or whitish, dry, firm, partial veil leaving hairs on the upper stalk or disappearing altogether, universal veil remnants leaving white fibrils or patches on the lower part of the stalk. **SPORES:** Rusty brown, elliptical, ornamented with low warts. **HABITAT:** Scattered on ground under conifers. **EDIBILITY:** Unknown but not recommended since some species of *Cortinarius* are deadly poisonous.

Cortinarius laniger

This is one of many difficult-to-distinguish species of *Cortinarius* with hygrophanous brown caps in the subgenus *Telamonia*. The color of the cap is a dull cinnamon brown at first but becomes a duller brown as the cap loses moisture. The stalk is also pale brown and is covered in part by white fibrils from the universal veil tissue. Although the odor of some collections is strongly of radishes, the odor of other collections is mild. Scores of other brownish species occur in this group, making field identification very difficult.

Cortinarius percomis

CAP: 4 to 12 cm in diameter, convex or hemispherical, yellowish to dull pale yellow-orange, aging buff, smooth, viscid when moist but soon dry, flesh yellow, odor sweet and distinctly fruity, taste mild. **GILLS:** Adnate to notched, dull yellow to buff becoming brown as the spores mature. **STALK:** 4 to 8 cm long, 1 to 2 cm wide, equal or clavate but lacking a basal bulb, pale yellow or dirty yellow, dry, firm, fibrillose, partial veil leaving a zone of

*Cortinarius
percomis*

hairs on the upper stalk that collects falling spores, universal veil yellow, fibrillose, leaving fibrils on the stalk. **SPORES:** Rusty brown, elliptical, covered with small warts. **HABITAT:** Single or scattered on ground under conifers. **EDIBILITY:** Unknown but not recommended since some species of *Cortinarius* are deadly poisonous.

The yellowish cap and stalk, yellow flesh, rusty brown spores, and especially the sweet fruity odor are the principal field marks. The cap is viscid in wet weather, but the stalk is dry. **Cortinarius albofragrans,** which also has a fruity odor, has a buff-colored cap, white flesh, whitish stalk, and preference for oaks. **Cortinarius papulosus** has a tan cap with a reddish brown center and bands of veil tissue on the stalk and lacks a distinctive odor. It is common under conifers.

Cortinarius ponderosus

CAP: 10 to 30 cm broad, convex or hemispherical, viscid when moist (but soon dry), ochre or yellow-brown with a yellow or ochre inrolled margin, the surface fibrillose and spotted with brown or reddish brown scales, becoming stained with rusty brown streaks, flesh very thick, yellowish, odor mild to sour. **GILLS:** Adnate to notched, purplish when young, becoming ochre and finally rusty brown when the spores mature. **STALK:** 8 to 20 cm long, 4 to 10 cm wide, equal or clavate, very firm, solid, viscid when moist, ochre, spotted with rusty brown fibrils and stains, partial veil disappearing or leaving a faint zone of hairs that collects falling spores. **SPORES:** Rusty brown, elliptical and ornamented with low warts. **HABITAT:** Single or scattered on ground in hardwood/coniferous forests. **EDIBILITY:** Unknown but not recommended since some species of *Cortinarius* are deadly poisonous.

This *Cortinarius* is characterized by its massive size, firm texture, and ochre-colored cap with rusty brown stains. **Cortinarius infractus** has a chestnut brown to dull brown, convex cap up to 15 cm in diameter. The stalk is clavate and whitish or light brown, and the flesh is buff or sometimes tinged violet, especially near the top of the stalk. The cap has a distinctly bitter taste, and the odor is mild to radishlike. It is common in coastal forests.

Cortinarius ponderosus

Cortinarius subfoetidus

Cortinarius subfoetidus

CAP: 3 to 10 cm in diameter, convex to plane, sometimes broadly umbonate, bright lavender to pinkish lilac, fading in age to bluish or paler, viscid to slimy, smooth, flesh tinged violet, odor fruity, like ripe pears. **GILLS:** Adnate to notched, lilac when young but becoming brown and finally rusty brown as the spores mature. **STALK:** 5 to 10 cm long, 1 to 2 cm wide, equal or slightly swollen near base, not viscid but usually moist, colored like cap or fading in age, smooth, partial veil leaving a faint zone of hairs on the stalk or disappearing, universal veil leaving a lavender sheath on the lower part of the stalk. **SPORES:** Rusty brown, elliptical, roughened with warts. **HABITAT:** Single, scattered, or in groups on ground in hardwood/coniferous forests. **EDIBILITY:** Unknown but not recommended since some species of *Cortinarius* are deadly poisonous.

This beautiful *Cortinarius* is recognized by its viscid, bright lavender cap, similarly colored stalk which is not viscid but may be moist, sweet odor, and rusty brown spores. The viscid cap separates it from the more frequently encountered **C. traganus** (p. 264), which has a lilac cap, reddish brown flesh, and gills that are never lilac or purplish, and **C. griseoviolaceus** (p. 258), which has a silvery lilac cap and stalk.

Cortinarius traganus

Cortinarius traganus

CAP: 4 to 12 cm in diameter, convex becoming plane, sometimes with a low umbo, lilac, smooth, dry or sticky, developing shallow cracks in age, sometimes partly covered with whitish or silver patches of veil tissue, the margin straight, flesh yellow or brown and mottled with reddish or yellow-brown streaks, odor fruity (like overripe pears). **GILLS:** Adnate to adnexed becoming notched, yellowish when young becoming rusty brown as the spores mature. **STALK:** 5 to 12 cm long, 1 to 3 cm wide, equal or wider at the base, solid, dry, smooth or silky, lilac, flesh reddish brown, partial veil lilac leaving a zone of hairs on the upper stalk, flesh mottled reddish or yellow-brown. **SPORES:** Rusty brown, elliptical, roughened from low warts. **HABITAT:** Single, scattered, or gregarious on ground under conifers. **EDIBILITY:** Unknown but not recommended since some species of *Cortinarius* are deadly poisonous.

The lilac color of the cap and stalk, rusty brown spores, gills that are never purplish, and reddish or yellowish brown interior are the distinctive field marks of this common mycorrhizal partner of conifers. **Cortinarius camphoratus** also has a lilac/violet cap and stalk, but it has an unpleasant odor of rotting potatoes and the interior of the stalk does not develop the brownish color found in *C. traganus*.

Cortinarius trivialis

CAP: 3 to 12 cm in diameter, convex to plane with or without a broad low umbo, grayish blue when young but soon becoming pale yellow-brown with olive hues, typically darker at the center, the blue colors disappearing altogether or remaining on the margin, viscid, smooth, flesh whitish, odor and taste mild. **GILLS:** Adnate or adnexed, grayish blue becoming rusty brown as the spores mature. **STALK:** 5 to 15 cm long, 1 to 2 cm wide, equal or slightly tapering downward, rigid, extending into the ground (rooting), viscid, white or pale yellow, the partial veil whitish, leaving a zone of hairs on the upper part of the stalk that becomes rusty brown from falling spores, universal veil leaving bands of yellow-brown fibrils and rings on the stalk. **SPORES:** Rusty brown, almond shaped, densely ornamented with relatively large warts. **HABITAT:** Single, scattered, or gregarious on ground broadleaf trees and shrubs, especially oak. **EDIBILITY:** Unknown but not recommended since some species of *Cortinarius* are deadly poisonous.

This species is a member of the *Myxacium* subgenus in the genus *Cortinarius*—those species with viscid caps and viscid stalks. The rusty brown spores, whitish stalk banded with yellow-brown rings, and association with oaks are diagnostic field marks. **Cortinarius collinitus** is similar, but it fruits in conifer-

Cortinarius trivialis

ous forests. It has a yellow- or orange-brown cap and white to
violet stalk banded with yellow-brown rings. **Cortinarius clidu-
chus** also has bands on the stalk, but the stalk is dry and the cap
is tawny.

Cortinarius vanduzerensis

CAP: 4 to 9 cm in diameter, conical when young and remaining
broadly conical or becoming convex-umbonate to plane, dark
brown or chestnut brown but fading in age, smooth, very viscid,
margin wavy at maturity, wrinkled, flesh whitish, odor mild.
GILLS: Adnexed or notched, buff when young becoming brown
or rusty brown. **STALK:** 8 to 20 cm long, 1 to 2 cm wide, equal or
gradually tapered downward, rooting into the substrate, viscid,
white or pale violet toward the apex, the rest covered with a
layer of purple slime, partial veil disappearing or leaving a few
hairs on the stalk. **SPORES:** Brown, elliptical, roughened from
low warts. **HABITAT:** Single or scattered on ground under conifers
in northern California coastal forests and farther north. **EDIBIL-
ITY:** Unknown but not recommended since some species of *Cortinarius* are deadly poisonous.

The conical to umbonate, deep brown or chestnut brown cap
covered with a thick layer of slime, violet-tinged slimy stalk, and
brown spores are the principal field marks. It is one of the easiest

Cortinarius vanduzerensis

species of *Cortinarius* to identify, primarily because of its copious slime layer. It might be confused with **C. collinitus,** which has an orange-brown viscid cap and stalk typically belted with veil remnants. Lilac tones may or may not be present. The cap is not as conical as *C. vanduzerensis*, at least when young.

Cortinarius verrucisporus

CAP: 3 to 7 cm in diameter, convex, yellow to yellow-brown, surface dry, flesh firm and yellowish, odor and taste mild. **GILLS:** Adnate to notched, white or cream colored to golden yellow when young, finally becoming brown as the spores mature. **STALK:** 1 to 3 cm long, 1 to 2 cm wide, equal or bulbous, colored like cap, solid, partial veil yellow, thin and silky, attached from the cap margin to the base of the stalk and not detaching, tearing radially in places to expose the gills. **SPORES:** Rusty brown, elliptical, ornamented with coarse warts. **HABITAT:** Usually in groups and clusters but sometimes occurring singly, buried or barely breaking through the soil or duff layer under conifers, often soon after the snow melts in the high mountains. **EDIBILITY:** Unknown but not recommended since some species of *Cortinarius* are deadly poisonous.

The yellowish cap, stocky stature, rusty brown spores, and especially the thin curtainlike partial veil that is persistently at-

Cortinarius verrucisporus

tached from the cap margin to the base of the stalk make this species easy to identify. ***Cortinarius magnivelatus*** is similar, but it has a tough white elastic veil. It also develops underground and usually remains buried or partially buried at maturity, fruiting under conifers in the mountains. The flesh of *C. magnivelatus* is white, not yellow like *C. verrucisporus*. It also has a longer, more developed stalk with a tapered base.

Cortinarius vibratilis

CAP: 2 to 5 cm in diameter, bell shaped, convex to plane with a broad low umbo, orange or brownish orange, fading in age, viscid, smooth, flesh white, odor mild, taste of the slime on the cap very bitter. **GILLS:** Adnate or notched, white becoming yellowish and finally rusty brown from the maturing spores. **STALK:** 5 to 8 cm long, up to 1 cm wide at the apex, typically clavate, viscid, white, smooth, surface sometimes bruising yellow, partial veil leaving a zone of hairs on the stalk that becomes brown from falling spores. **SPORES:** Rusty brown, elliptical, covered with low warts. **HABITAT:** Scattered or in small groups in hardwood/coniferous forests. **EDIBILITY:** Unknown but not recommended since some species of *Cortinarius* are deadly poisonous.

The orange viscid cap that contrasts sharply with the white stalk, rusty brown spores, and bitter taste of the cap's slimy layer,

Cortinarius vibratilis

which is readily detected, characterize this species. **Cortinarius mucosus** is similar but is typically larger. It has a viscid orange-brown cap 4 to 10 cm broad and a viscid stalk 5 to 15 cm long and 1 to 3 cm wide. In addition, the cap of *C. mucosus* is not bitter tasting, and the stalk is more-or-less equal rather than clavate. *Cortinarius mucosus* is widespread but is particularly common under conifers in the mountains of the West.

Cortinarius violaceus
Violet Web Cap

CAP: 5 to 12 cm in diameter, convex to plane, sometimes with a low umbo, dark violet, dry, covered with tufts of short erect fibers, flesh violet with white streaks, odor cedarlike. **GILLS:** Adnexed or notched, close, dark violet becoming brown from the maturing spores. **STALK:** 6 to 18 cm long, 1 to 2 cm wide above, equal or somewhat clavate, fibrillose, colored like the cap, firm, partial veil grayish purple, leaving a cobweblike zone of hairs that collects falling spores. **SPORES:** Rusty brown, elliptical, covered with low warts. **HABITAT:** Single or in small groups on ground under conifers. **EDIBILITY:** Unknown but not recommended since some species of *Cortinarius* are deadly poisonous.

The deep violet color (almost black), tufts of short fibers that cover the surface of the cap, and rusty brown spores make this

Cortinarius violaceus

an easy species to identify. It is one of the darkest of all *Cortinarius* species. Other dark violet or blackish mushrooms include **Entoloma parvum** and relatives (p. 200), but those are small, and **E. bloxami** (p. 195), which has pinkish spores. The cap of **Cortinarius cotoneus** (p. 254) is also densely covered with tufts of short hairs but is olive brown to yellow-brown with a darker center.

Cortinarius xanthodryophilus

CAP: 6 to 10 cm in diameter, convex, becoming plane to uplifted in age, yellow when young, aging yellow-brown, frequently with reddish brown discolorations, the margin typically inrolled well into maturity, surface viscid when moist, becoming dull glossy when dry, the universal veil leaving white patches on the surface, flesh mostly white but tinged blue above the gills, odor and taste mild. **GILLS:** Notched, crowded, pale yellow, becoming brown as the spores mature. **STALK:** 5 to 10 cm long, 1.5 to 3 cm wide with a large rimmed bulb at the base up to 5 cm broad, dry, white, occasionally with blue tints on the upper part, the flesh sometimes tinged blue, the partial veil usually leaving a zone of long hairs that soon becomes brown from falling spores. **SPORES:** Rusty brown, elliptical to almond shaped, coarsely roughened with warts. **HABITAT:** Single to gregarious on ground under oaks and tanoaks. **EDIBILITY:** Unknown but not recommended since some species of *Cortinarius* are deadly poisonous.

This species is recognized by the yellowish color of the viscid cap that is frequently decorated with whitish patches, pale colored gills when young, rusty brown spores, and large rimmed basal bulb. The viscid cap and dry stalk broadly characterize it as a member of the subgenus *Phlegmacium*. Several other related yellowish species with basal bulbs (loosely called bulbopodiums) are encountered under oaks and pines. Some lack any hints of blue. Many have not been formally named.

Cantharellus californicus
Chanterelle

CAP: 5 to 30 cm in diameter, plane to depressed, yellow-orange to bright orange, the surface smooth, the margin often wavy, flesh thick, pale yellow, odor mild to fruity, taste mild. **SPORE-**

Cortinarius xanthodryophilus

BEARING SURFACE: Decurrent shallow folds and wrinkles, forked and veined, colored like cap or with a pinkish tinge. **STALK:** 2 to 8 cm long 1 to 4 cm wide, equal or tapering toward the base, yellow, discoloring darker, dry, smooth, firm. **SPORES:** Creamy or pale yellow, elliptical, smooth. **HABITAT:** Single, scattered, or gre-

Cantharellus californicus

garious on the ground in mycorrhizal association with oaks. **ED-IBILITY:** Edible and choice (and usually bug free).

This highly regarded mushroom is identified by its yellow-orange color, ridged (rather than bladelike) gills that are often veined, forked, or both, and meaty texture. It often has a pleasant apricot odor, at least when fresh. The West has several similar but smaller species: **C. formosus,** a dull orange species found under conifers and tanoak that has a cap with a smooth to suedelike surface, spore-bearing surface sometimes tinged pink, and a relatively thin stalk; **C. cascadensis,** bright yellow with a clavate to bulbous stalk, found under Douglas-fir in the Pacific Northwest; and **C. cibarius var. roseocanus,** which is distinguished by an association with spruce, a pinkish bloom on the yellow-orange cap, and bright golden orange fertile surface. All are edible.

Cantharellus subalbidus
White Chanterelle

CAP: 5 to 15 cm in diameter, plane to depressed becoming vase shaped in age, white to cream colored, often with yellow-brown stains, smooth or developing shallow cracks, firm, the margin often wavy or irregular, flesh thick, white, odor and taste mild.

Cantharellus subalbidus

SPORE-BEARING SURFACE: Decurrent shallow folds and wrinkles, forked and veined, colored like the cap or paler. **STALK:** 2 to 7 cm long 1 to 5 cm wide, tapering toward the base, colored like the cap, dry, smooth, firm, developing yellow-brown spots where bruised or in age. **SPORES:** White, elliptical, smooth. **HABITAT:** Single, scattered, or gregarious on ground in mycorrhizal association with trees and shrubs in hardwood/coniferous forests. **EDIBILITY:** Edible and choice (and usually bug free).

The White Chanterelle is similar to the better known yellow-orange Chanterelle, *Cantharellus californicus,* in all respects except that it is white to cream colored and lacks the apricot odor of the Chanterelle. Like the Chanterelle, the spores are born on foldlike wrinkles and ridges instead of bladelike gills. It is not likely to be confused with any other mushroom. **Leucopaxillus albissimus** (p. 138), which is also uniformly white, has platelike gills and a dense mycelial mat at the base of the stalk.

Craterellus cornucopioides
Black Trumpet, Horn of Plenty

CAP: 2 to 7 cm broad, fruiting body up to 12 cm tall, funnel or vase shaped and hollow in the center (tubular), dark gray to black (or very rarely yellow), the margin rolled back (like fused petals of a flower) and wavy, inconspicuously scaly, dry, thin, leathery, odor and taste mild. **SPORE-BEARING SURFACE:** Smooth

Craterellus cornucopioides

or with very shallow decurrent wrinkles, pale gray. **STALK:** 1 to 5 cm long, 0.5 to 1 cm wide, continuous with cap, tapering to the base, black, hollow to the point of attachment with the ground. **SPORES:** Whitish to pale yellow, elliptical, smooth. **HABITAT:** Scattered, gregarious, or often in clusters on ground in hardwood/coniferous forests. **EDIBILITY:** Edible and choice.

The blackish coloration, tubular form, and smooth or faintly wrinkled undersurface are principal field marks. Fruiting bodies commonly occur in clusters. An individual fruiting body is quite thin and light. Their dark color hides them well on the forest floor, but when one is found, more are likely nearby. The Black Chanterelle, *Craterellus cinereus*, is similarly colored, but although funnel shaped, it is not tubular like *C. cornucopioides*. It has distinct folds and wrinkles on the underside of the cap and a fragrant fruity odor.

Craterellus tubaeformis

(Other name: *Cantharellus infundibuliformis*) Winter Chanterelle, Yellow Foot

CAP: 1 to 4 cm in diameter, plane with a depressed center, funnel or trumpet shaped (the center sometimes hollow), yellow to yellow-brown, fading in age, waxy, the margin incurved and wavy, flesh thin and yellow-brown, odor and taste mild. **SPORE-BEARING SURFACE:** Decurrent shallow, blunt, thick, forked gills with cross-veins, yellow with orange or brown tones but generally paler than the cap. **STALK:** 2 to 8 cm long, up to 1 cm wide, equal or tapering toward the base, colored like the cap, often longitudinally grooved, hollow in age. **SPORES:** White to buff, elliptical, smooth. **HABITAT:** Scattered or gregarious on ground, moss, or rotten wood under conifers, especially common late in the season. **EDIBILITY:** Edible; it is sometimes sold in markets as a Chanterelle, but it does not compare with *Cantharellus* species in flavor or substance.

This small chanterelle is yellow-brown with decurrent blunt gills. The fruiting body often becomes hollow in age, giving rise to another common name, Tubies. It is generally duller in color than the bright golden orange of the more robust **Cantharellus californicus** (p. 270), and it typically fruits later in the mushroom season, often in the company of **Hydnum repandum** (p. 283), one of the tooth fungi.

Craterellus tubaeformis

Gomphus clavatus
Pig's Ears

CAP: Up to 18 cm tall and broad, vase shaped, plane or centrally depressed, tan to light brown and sometimes with lilac tones, fading paler in age, the margin wavy, lobed, and often uplifted in age, firm, the surface smooth or cracked when dry, flesh whitish

Gomphus clavatus

or tinged lilac, odor and taste mild. **SPORE-BEARING SURFACE:**
Folds and wrinkles, decurrent, purple or purplish tan, fading to
tan in age. **STALK:** 3 to 6 cm long, 1 to 3 cm wide, firm, solid,
central or eccentric, continuous with the cap, tapering toward
the base, usually supporting multiple caps, lilac-tan above, whitish toward the base. **SPORES:** Yellowish, elliptical, wrinkled or
finely warted, inamyloid. **HABITAT:** Scattered or growing in clusters fused together on ground in mycorrhizal association with
conifers. **EDIBILITY:** Edible; some sources consider it choice,
whereas others describe its flavor as mediocre to nonexistent.
When mature, the clusters are often buggy.

The purplish fertile surface composed of folds and wrinkles
rather than the typical bladelike gills of most mushrooms, tan
cap, and tendency to grow in fused groups make this species
easy to identify. Compare it to the equally common *Turbinellus
floccosus* (below), which also has folds and wrinkles on the undersurface but is bright orange.

Turbinellus floccosus
(Other name: *Gomphus floccosus*) Scaly Chanterelle
CAP: 5 to 15 cm in diameter and up to 20 cm or more tall, vase
shaped, deeply depressed in the center or hollow, surface bright
reddish orange fading to dull orange with large darker scales,

Turbinellus floccosus

the margin wavy, flesh white, odor and taste mild. **SPORE-BEARING SURFACE:** Creamy yellow, yellow, to buff, decurrent folds and wrinkles. **STALK:** 3 to 10 cm tall, 1 to 3 cm wide, continuous with the cap, central or slightly eccentric, tapering toward the base, becoming hollow in age, yellowish to buff, smooth. **SPORES:** Yellowish, elliptical, wrinkled or warted, inamyloid. **HABITAT:** Single, scattered, or in clusters on ground under conifers. **EDIBILITY:** Not recommended; edible for some, causing gastrointestinal upsets in others.

This tall, scaly, bright reddish orange, vase-shaped mushroom is eye-catching. The fertile surface consists of folds that run down the stalk and are continuous with the cap. The center of the fruiting body becomes hollow in age. It could only be confused with a relative, **T. kauffmanii** (*Gomphus kauffmanii*), which is buff to brown in color. The common name, Scaly Chanterelle, is based on superficial similarity to true chanterelles; DNA evidence shows that *T. floccosus* is more closely related to stinkhorns and some coral fungi than to chanterelles. Superficially, *T. floccosus* is morphologically similar to **Gomphus clavatus** (p. 275), but based on genetic evidence, it is distantly related.

Tooth Fungi

This is an artificial group united only by the presence of teeth or spines, which bear the spores. Some have caps and stalks, whereas others are cushion shaped on wood. Species that grow on the ground with a central stalk are usually mycorrhizal; those that grow on wood or conifer cones are saprobic. *Hericium erinaceus*, which is easily cultivated, and *Hydnum repandum* are choice edibles; others are inedible due to their tough texture or bitter taste. Some species of *Hydnellum* and *Sarcodon* are used to produce natural dyes. One jelly fungus, *Pseudohydnum gelatinosum*, is included here because it has tiny spinelike projections on the undersurface of the cap. It has a rubbery texture unlike the firm or fleshy texture of the tooth fungi included here.

a. Growing on wood or conifer cones
  b. Fruiting body rubbery, translucent white.............................
    ......**Pseudohydnum gelatinosum** (a jelly fungus, see p. 308)
  b. Fruiting body not rubbery or translucent

     c. Fruiting body a mass of white icicle-like spines; on wood
.... **Hericium erinaceus**

     c. Cap kidney shaped, hairy, brown; on decaying conifer
cones .... **Auriscalpium vulgare**

a. Growing on the ground

   d. Cap surface lumpy, sometimes encompassing needles and
leaves as it grows

     e. Cap surface orange; flesh orange...
.... **Hydnellum aurantiacum**

     e. Cap white and then brown, beaded with bright red
drops .... **Hydnellum peckii**

   d. Cap fleshy or leathery, not encompassing needles and
leaves as it grows

     f. Cap pale orange, smooth.................... **Hydnum repandum**

     f. Cap not pale orange

       g. Cap tough and fibrous, surface smooth or ridged........
.... **Phellodon tomentosus**

       g. Cap fleshy, surface scaly

         h. Cap vinaceous brown....... **Sarcodon subincarnatus**

         h. Cap brown with prominent scales.............................
.... **Sarcodon imbricatus**

The spore-bearing teeth on the undersurface of the cap of *Hydnellum aurantiacum*

Auriscalpium vulgare
Ear Pick Mushroom

CAP: 1 to 3 cm broad, kidney shaped, broadly convex to plane, dark reddish brown with a light brown margin, covered with reddish brown stiff hairs, dry, leathery and tough, odor mild. **SPORE-BEARING SURFACE:** Small spines hanging from the underside of the cap, 1 to 3 mm long, whitish, violet, or pinkish brown. **STALK:** 2 to 8 cm long, 1 to 3 mm wide, equal or slightly enlarged at the base, attached to the side of the cap, dark reddish brown, hairy, tough. **SPORES:** White, subglobose, roughened, amyloid. **HABITAT:** Single or in small numbers on decaying Douglas-fir or pine cones. **EDIBILITY:** Unknown, but too small to be of value.

The small size of the fruiting body, spines on the undersurface of the hairy reddish brown cap, and growth on conifer cones make this an easy mushroom to identify. Because of its small size, however, it is often overlooked in the dark of the forest floor. It seems to prefer Douglas-fir cones in the West. Other cone inhabitants include **Strobilurus trullisatus** (p. 172), which also fruits on Douglas-fir cones, **Baeospora myosura** on spruce cones, and **Mycena purpureofusca** (p. 182), which often fruits on pine cones. All of theses species have typical platelike gills as their spore-bearing surface rather than spines like *A. vulgare.*

Auriscalpium vulgare

Hericium erinaceus

Hericium erinaceus

Lion's Mane

FRUITING BODY: A fleshy mass of white, unbranched, soft, icicle-like spines; a single fruiting body, which is circular to slightly elongate, may grow to over 30 cm in diameter, although specimens 15 to 20 cm are typical. **SPORE-BEARING SURFACE:** Spines which are short when young and up to 5 cm or more in length when mature; in age, the spines discolor a dirty yellow. **STALK:** Absent. **SPORES:** White, globose, roughened with minute warts. **HABITAT:** Generally solitary on dead hardwoods or on wounds of living trees. **EDIBILITY:** Edible and choice.

The "icicles" or soft spines on a white cushion growing on wood are infallible field marks. It has no poisonous look-alikes. Fruiting bodies of this mushroom are likely to be found in the same location year after year. It is often sold in Asian markets for both its food and medicinal value. ***Hericium coralloides*** is similar, but it has short white spines arranged in rows and an open growth habit. It lacks the central cushion of *H. erinaceus*. *Hericium erinaceus* is easily cultivated on sawdust and is sold under the names Lion's Mane, Monkey's Head, and Old Man's Beard, among others.

Hydnellum aurantiacum

CAP: 3 to 15 cm broad but sometimes fusing with others and appearing larger, encompassing twigs and needles as it grows, rough and knobby but more-or-less circular in outline, plane or depressed in the center, surface velvety, orange to rusty orange, becoming dark brown in age, the actively growing margin paler, flesh tough, orange, odor mild to fragrant, taste bitter. **SPORE-BEARING SURFACE:** Short crowded spines 1 to 4 cm long, white to orange, brown in age (p. 278). **STALK:** 2 to 5 cm long, 1 to 3 cm wide, more-or-less central, tough, equal or irregularly shaped, orange to rusty brown becoming brown in age, incorporating plant debris as it grows. **SPORES:** Brown, subglobose, warted. **HABITAT:** Single or in small groups on ground under conifers, especially pines. **EDIBILITY:** Inedible.

The orange color of the cap and flesh, knobby cap surface, and presence of spines are the principal characteristics of this species. **Hydnellum caeruleum** is similar, but the caps have bluish tints. In age, the cap loses the blue color and becomes brown. However, bluish tints remain in the flesh of the cap, which is faintly zoned. The interior of the stalk is rusty orange. It grows on the ground under both conifers and hardwoods. **Hydnellum suaveolens** has a yellowish brown cap, a violet interior of both the cap and stalk, and a very fragrant aniselike odor.

Hydnellum aurantiacum

Hydnellum peckii
Strawberries and Cream

CAP: 3 to 15 cm broad, often with embedded twigs and needles that it surrounds and encompasses as it grows, plane or centrally depressed, becoming lumpy and uneven but more-or-less circular in outline, velvety at first, white when young but becoming brown from the center toward the margin in age, the actively growing margin white or pink, aging brown, exuding droplets of red juice on the surface of the cap, flesh reddish brown, odor mild or unpleasant, taste acrid. **SPORE-BEARING SURFACE:** Short crowded spines 1 to 6 mm long, pink becoming brown, often running down the stalk. **STALK:** 2 to 7 cm long, 1 to 2 cm wide, central or off-center, more-or-less equal, velvety, brown, tough. **SPORES:** Brown, subglobose, ornamented with warts. **HABITAT:** Single, scattered, or gregarious on ground under conifers. **EDIBILITY:** Inedible, the texture is tough and the taste is acrid (peppery).

When actively growing, the bright red droplets on the white growing surface make this mushroom easy to identify. In age, the cap darkens to brown and field identification is difficult. See **H. aurantiacum** (p. 281) for other tooth fungi with lumpy cap surfaces in different hues.

Hydnellum peckii

Hydnum repandum

Hydnum repandum

(Other name: *Dentinum repandum*) Hedgehog Mushroom

CAP: 2 to 12 cm in diameter, convex but soon plane, the margin often wavy, cream colored to dull orange, aging tan or reddish tan, the surface dry, smooth, unpolished, flesh buff, brittle, bruising brownish orange, odor and taste mild to slightly peppery. **SPORE-BEARING SURFACE:** Spines 3 to 6 mm long, adnate or sometimes decurrent, cream colored. **STALK:** 3 to 8 cm long, 1 to 3 cm wide, central or eccentric, equal or enlarged at the base, cream colored to buff, bruising brownish orange, dry, smooth, partial veil absent. **SPORES:** White, subglobose, smooth. **HABITAT:** Single, scattered, or gregarious on ground in hardwood/coniferous forests, especially common under conifers. **EDIBILITY:** Edible and choice.

This much sought-after mushroom is easily recognized by its creamy to dull orange color and pale spines. The cap of the similar **H. umbilicatum** (*Dentinum umbilicatum*) is 3 to 5 cm in diameter, convex to plane, and often umbilicate (with a deep central depression). The stalk is usually central, 3 to 6 cm long, and 5 to 10 mm wide, which is considerably thinner than the stalk of *H. repandum*. It also grows on the ground under conifers. Along the California coast, these mushrooms tend to fruit after the peak of the mushroom season.

Phellodon tomentosus

Phellodon tomentosus

CAP: 1.5 to 5 cm broad, plane or funnel shaped, often fused with others, concentrically zoned with yellow-brown, reddish brown, and brown colors, the growing margin whitish, surface dry, velvety, smooth or ridged, flesh thin, brown and leathery, odor fragrant, taste mild or bitter. **SPORE-BEARING SURFACE:** Short crowded spines 1 to 3 mm long, grayish brown with paler tips, running down the stalk a short distance. **STALK:** 1 to 5 cm long, 2 to 5 mm wide, more-or-less central, equal, colored like cap. **SPORES:** White, globose or subglobose, spiny. **HABITAT:** Scattered, gregarious, or clustered on ground under conifers. **EDIBILITY:** Inedible.

The zoned yellow-brown to brown thin cap, brown flesh, and presence of spines are field marks of this small tooth fungus. **Phellodon atratus** is similar but has a purplish black, concentrically zoned cap with a whitish or pale purple growing margin and purplish gray spines. *Hydnellum* species may also have brownish zoned caps, but the caps are larger and thicker, and they have brown, not white, spores. *Phellodon* species can be confused with **Coltricia cinnamomea,** but that species has pores rather than spines. It has a thin, cinnamon, fibrillose, zoned cap

(1 to 5 cm broad) with brown flesh and a rusty brown stalk. **Sarcodon fuscoindicus** (*Hydnum fuscoindicum*) is a much larger violet mushroom with violet spines that run down the stalk and brown spores.

Sarcodon imbricatus

(Other name: *Hydnum imbricatum*) Shingled Hedgehog

CAP: 5 to 20 cm in diameter, convex to plane with a depressed center, dry, tan to reddish brown with conspicuous large, dark brown, shinglelike, upturned scales, the margin inrolled at first, flesh cream colored to pale brown, odor mild, taste mild to bitter. **SPORE-BEARING SURFACE:** Spines 5 to 15 mm long, adnate to slightly decurrent, light brown or grayish brown becoming dark brown in age. **STALK:** 4 to 10 cm long, 2 to 4 cm wide, central or eccentric, enlarged at the base, tan or brown, dry, smooth, partial veil absent. **SPORES:** Brown, elliptical to subglobose, ornamented with warts. **HABITAT:** Single, scattered, or gregarious on ground in hardwood/coniferous forests. **EDIBILITY:** Edible but not recommended since it causes gastrointestinal upsets in many people.

The large brown scales on this robust mushroom and spines instead of gills are defining characteristics. It is especially common in the Rocky Mountains after summer rain but is wide-

Sarcodon imbricatus

spread in other western states. **Sarcodon scabrosus** (*Hydnum scabrosum*) has a smoother cap when young and conspicuous scales in age that either do not become erect or are limited to the center of the cap, a stalk with a dark green to blackish base, a mild to mealy odor, and bitter taste. The flesh has a lavender tint.

Sarcodon subincarnatus

CAP: Up to 15 cm broad, convex, plane, or centrally depressed, vinaceous brown, dry, smooth when young but scaly in age, bruising brown, flesh thick, brittle, and firm, pale vinaceous colored, odor mealy, taste bitter. **SPORE-BEARING SURFACE:** Pale vinaceous-violet spines with paler tips, soft, 2 to 6 mm long, decurrent. **STALK:** 2 to 10 cm long, 2 to 6 cm wide, equal or tapered toward the base, central or eccentric, colored like cap but paler. **SPORES:** Brown, globose to subglobose, ornamented with warts. **HABITAT:** Single or scattered on ground in hardwood/coniferous forests. **EDIBILITY:** Inedible due to the bitter taste.

This mushroom is recognized by the vinaceous-brown color of the cap, pale flesh, and spore-bearing spines that extend down the stalk. **Sarcodon fuscoindicus**, which is used to produce a blue dye, has a violet-black cap with a paler margin, violet flesh, and dark violet spines. It resembles **Phellodon atratus**, but that

Sarcodon subincarnatus

species is smaller, thinner, and white spored and has a tougher texture. **Sarcodon fuligineoviolaceus** (*Hydnum fuligineoviolaceum*) has a reddish brown cap, brownish spines, gray interior of the base of the stalk, and an acrid taste. Both **S. imbricatus** (p. 285) and **S. scabrosus** have large brown scales.

Sparassis radicata
(Other name: *Sparassis crispa*) Cauliflower Mushroom

FRUITING BODY: A mass of flattened, ribbonlike, wavy and curly branches (like egg noodles), the entire fruiting body usually more-or-less spherical in outline, but this is highly variable, 12 to 40 cm or more in diameter, branches smooth, cream colored becoming tan and brown edged in age. **STALK:** 5 to 12 cm or more long, 2 to 5 cm wide, rootlike, flesh white. **SPORES:** White, elliptical, smooth, inamyloid. **HABITAT:** Fruiting at or near the base of conifers, growing from the roots. **EDIBILITY:** Edible and choice, stores well in the refrigerator.

The mass of yellowish, ribbonlike wavy branches and the growth at the base of tree trunks make this mushroom easy to recognize. It often fruits for several years from the same tree, causing a brown rot disease of the major roots of its host. Typi-

Sparassis radicata

cally, only one fruiting body is produced from an individual tree. Differences in the structure of the stalk and in DNA sequences have caused researchers to separate *S. radicata* and **S. crispa,** but the distinction has not been adopted by everyone. Surprisingly, *Sparassis* species are closely related to *Phaeolus, Laetiporus, Pycnoporellus, Oligoporus,* and other polypores that cause brown rot of trees.

Club and Coral Fungi

This group is a loose association of fungi united here by outward appearance and not necessarily by ancestry. The fruiting bodies are erect and clublike or profusely branched and coral-like. Spores are borne on the outer surfaces of the branches. Microscopic features of the hyphae and spores separate the various genera. Club and coral fungi are not widely harvested for food; some are edible, but others cause gastrointestinal upsets. Some live on decaying forest litter, but most are probably mycorrhizal with various trees and shrubs. In addition to the species described here, see *Calocera cornea* (p. 304) a small, orange, erect jelly fungus, and *Xylaria hypoxylon* (p. 380), a black, antlerlike ascomycete. Both are coral-like in stature but unrelated to any of the club and coral fungi.

a. Fruiting body unbranched
    b. Apex often enlarged; fruiting body not fragile .... ***Clavariadelphus occidentalis***
    b. Apex not enlarged; fruiting body fragile
        c. Fruiting body white....................................***Clavaria fragilis***
        c. Fruiting body orange ....***Clavulinopsis laeticolor***
        c. Fruiting body brown.......................***Macrotyphula juncea***

a. Fruiting body branched
    d. Typically less than 8 cm tall; stalk slender
        e. Branches white, the ends rounded and flattened ....***Tremellodendropsis tuberosa***
        e. Branches white or yellowish; the tips ending in sharp or dull points
            f. Branches white ....***Ramariopsis kunzei***
            f. Branches yellow to ochre ....***Ramaria myceliosa***
    d. Typically more than 8 cm tall; stalk usually thick

g. Fruiting body entirely magenta or coral red.....................
.... ***Ramaria araiospora***

g. Fruiting body not predominately red, but branch tips may be reddish

   h. Fruiting body whitish or yellowish with reddish branch tips or discoloring reddish somewhere on branches, stalk, or both

      i. Fruiting body whitish with vinaceous tips; very compact...***Ramaria botrytis***

      i. Fruiting body cream colored with reddish stains on the lower branches....................***Ramaria rubiginosa***

   h. Fruiting body predominately yellow, orange, or brown

      j. Branches pinkish yellow or pinkish orange, tips yellow ....***Ramaria formosa***

      j. Fruiting body otherwise colored

         k. Branches predominately orange ....
....***Ramaria sandaracina***

         k. Branches predominately brown...........................
....................................***Ramaria violaceibrunnea***

         k. Branches predominately yellow...........................
....***Ramaria rasilispora***

Clavaria fragilis

(Other name: *Clavaria vermicularis*) Fairy Fingers

FRUITING BODY: Slender, unbranched, erect, white "fingers," 3 to 10 cm tall, 2 to 4 mm wide, cylindrical or flattened in cross section, tapered toward the tip, smooth, fragile, and brittle, likely to break when picked. **SPORES:** White, elliptical, smooth. **HABITAT:** Typically in clusters or tufts but fruiting singly as well on the ground in forests. **EDIBILITY:** Edible but too small to be of value.

Clavaria fragilis is easily recognized by the white, thin, and fragile fingers. It usually grows in clusters. Other unbranched club fungi include **Alloclavaria purpurea** (*Clavaria purpurea*), purple to purplish brown and also fruiting in clusters, and **Macrotyphula juncea** (p. 292), an extremely slender (less than 2 mm wide) brown club fungus that grows scattered on the ground. Because of its small size and colors that blend into the forest floor, it is difficult to spot, but it sometimes fruits in large numbers. More conspicuous is **Clavulinopsis laeticolor** (p. 291),

Clavaria fragilis

bright orange to yellow-orange fingers 2 to 5 cm tall, 1 to 3 mm wide, and sometimes flattened in cross section. It grows singly or in loose tufts. **Clavulinopsis fusiformis** (*Ramariopsis fusiformis*) is similar but yellow. It often grows in bundles that are fused at the base.

Clavariadelphus occidentalis

(Other name: *Clavariadelphus pistillaris*) Club Fungus

FRUITING BODY: Unbranched or occasionally forked club, up to 15 cm tall, 1 to 3 cm wide at the top and tapering toward the base, smooth to wrinkled, pinkish brown to yellow-brown, the flesh white, staining reddish brown when cut, odor mild, taste often bitter. **SPORES:** White, elliptical, smooth. **HABITAT:** Scattered to gregarious on ground in mycorrhizal association with conifers. **EDIBILITY:** The unpleasant taste is a deterrent.

Club fungi are readily identified by their shape; *C. occidentalis* is club shaped with a rounded apex and some shade of brown,

Clavariadelphus occidentalis

whereas **C. truncatus** is brownish orange to pinkish brown with a bright yellow flattened top. It is edible, sweet, and considered the best of the club fungi for the table. When fresh, it can be eaten raw in salads or even used in desserts; older specimens lose both sweetness and crunchiness. **Clavariadelphus ligula** is yellowish and has a rounded top. It often fruits in large numbers, sometimes carpeting the forest floor in western mountains. Another fairly frequently encountered club fungus is **C. caespitosus,** which is grayish red or cinnamon buff and grows in clusters.

Clavulinopsis laeticolor
(Other name: *Ramariopsis laeticolor*)
FRUITING BODY: Slender, unbranched, erect, bright orange or yellow fingers, less than 5 cm tall and 3 mm wide, fading to a duller yellow in age, cylindrical and sometimes grooved or flattened in cross section, tapered toward the rounded tip, smooth, dry, flesh yellow and thin, odor and taste mild. **SPORES:** White, elliptical to almost round, smooth. **HABITAT:** Single, scattered, or in loose groups on the ground under hardwoods and conifers. **EDIBILITY:** Unknown but too small to be of value.

Despite its small size, this club fungus is easy to spot in the darkness of the forest. True to its species name, it is brightly col-

Clavulinopsis laeticolor

ored. **Clavulinopsis fusiformis** (*Ramariopsis fusiformis*) is similar. It is bright or pale yellow to orange, unbranched, cylindrical, and up to 15 cm tall. It is distinguishable from *C. laeticolor* by its larger size, bitter taste, and habit of growing in tight bundles fused at the base. It often has pointed tips. Other slender club fungi include white **Clavaria fragilis** (p. 289), brown and extremely thin **Macrotyphula juncea** (below), and purplish **Alloclavaria purpurea** (*Clavaria purpurea*). The jelly fungus **Calocera cornea** (p. 304) has erect, orange, cylindrical fruiting bodies up to 1.5 cm high. It fruits on wood.

Macrotyphula juncea

FRUITING BODY: Very thin and threadlike, 3 to 8 cm tall overall and less than 2 mm wide, unbranched, erect, yellowish brown to brown, cylindrical or tapering upward, smooth, the tip pointed or blunt, the stalk slightly narrower than the fertile upper part and sparsely hairy near the base, flesh firm, odor and taste mild. **SPORES:** White, elliptical to teardrop shaped, smooth. **HABITAT:** Scattered or gregarious in conifer and hardwood leaf litter and rotting twigs, including Coast Redwood litter, sometimes fruiting in large numbers. **EDIBILITY:** Unknown but much too small to be of interest.

Macrotyphula juncea

At first glance, *M. juncea* looks like tiny bare stems of a small plant growing out of the forest floor. Each fruiting body is thin, erect, brown, and usually unbranched. Other unbranched club fungi include **Clavaria fragilis** (p. 289), which has white finger-like fruiting bodies that are more robust (up to 4 mm wide) and usually grows in clusters, unlike *M. juncea*, which grows scattered on the ground; **Alloclavaria purpurea** (*Clavaria purpurea*), which is purplish, up to 6 mm wide, and fruits in clusters; and **Clavulinopsis laeticolor** (p. 291), bright orange to yellow-orange, up to 3 mm wide, and growing singly or in loose tufts. **Xylaria hypoxylon** (p. 380) is white and black or completely black, branched or unbranched, and grows on decaying wood.

Ramariopsis kunzei

FRUITING BODY: Coral-like, up to 10 cm high and 8 cm broad, repeatedly and dichotomously branched, the forks of the branches usually U-shaped, white or sometimes cream colored in age, tips pointed or rounded, surface smooth, texture fragile and brittle, odor and taste mild. **STALK:** Absent, or if present, very short and minutely hairy. **SPORES:** White, subglobose to elliptical, warted, inamyloid, produced on four-spored basidia. **HABITAT:** Single, scattered, or gregarious on ground in hardwood/coniferous forests. **EDIBILITY:** Edible but insubstantial.

Ramariopsis kunzei

The white color of the fruiting body, moderate size, and U-shaped pattern of branching are the important field marks. Several other small to moderately sized coral-like fungi are common in forests. **Clavulina cristata** is also uniformly white and has flattened, finely toothed branch tips, **C. cinerea** is similar but is lilac-gray, and like *C. cristata*, it has two-spored basidia, and **C. rugosa** has wrinkled white branches and is sparsely branched. **Artomyces pyxidatus** (*Clavicorona pyxidata*) has pale yellow to brownish crownlike branch tips and grows on fallen hardwoods (see pp. 44–45), and **Clavulinopsis corniculata** (*Ramariopsis corniculata*) is bright yellow to ochre, sparingly branched, and up to 5 cm broad and grows on the ground in woods, in grassy areas, or on well-decayed wood. All of these small coral-like fungi have white spores.

Ramaria araiospora

FRUITING BODY: Coral-like, 5 to 14 cm high, 3 to 10 cm wide, repeatedly branched, branches bright magenta or coral red, the tips similarly colored or becoming yellow to orange, flesh colored like branches, brittle, odor and taste mild. **STALK:** 2 to 3 cm long or longer, up to 2 cm wide, white at the base. **SPORES:** Yellowish, elliptical to cylindrical, roughened with tiny warts. **HABITAT:** Single or scattered on ground in mycorrhizal association

Ramaria araiospora

with hardwoods and conifers. **EDIBILITY:** Edible, but like many coral fungi, it should be tried with caution since some corals have a laxative effect on certain people.

The bright magenta color makes *R. araiospora* one of the most distinctive and beautiful of all the coral fungi. With age, the red color fades and the fruiting body may become a dull yellow, like many coral fungi, but a hint of red usually remains. ***Ramaria stuntzii*** is similar but has a larger base and reddish orange lower branches, and the flesh is amyloid (the flesh of *R. araiospora* is inamyloid). Some specimens of *R. formosa* are pink, but the tips of the branches are usually yellow, and it never has the deep red color of *R. araiospora*.

Ramaria botrytis

FRUITING BODY: Coral-like, 7 to 20 cm high and 6 to 30 cm broad, profusely, densely, and compactly branched, the basal branches thick and whitish, compact when young, terminal branches short and crowded, white with pink, purple or vinaceous tips with two to four points, odor mild or pleasant, taste mild. **STALK:** 3 to 5 cm long, 2 to 5 cm wide, often tapering toward the base,

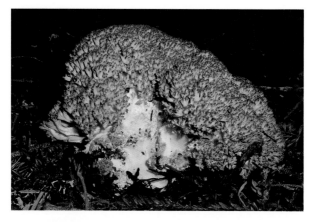

Ramaria botrytis

whitish, firm, solid. **SPORES:** Yellowish, long-elliptical, longitudinally striate. **HABITAT:** Single or scattered or in arcs and rings on ground in mycorrhizal association with conifers and hardwoods. **EDIBILITY:** Edible, although laxative effects have been reported.

This is one of the easiest coral fungi to identify. The purplish or vinaceous tips of the otherwise whitish branches, compact stature, and the cauliflower-like look are the principal field marks of this large and common species. In age, the color of the branches and the tips fade to brown and the fruiting body loses the compact appearance; however, even in old age lower branches may reveal telltale vinaceous tips. *Ramaria rubrievanescens* has pink on the branch tips only when young; in age it becomes entirely cream colored.

Ramaria formosa

FRUITING BODY: Coral-like, 12 to 20 cm high and broad, branching repeatedly and often, the branches vertical at least when young, branches thick at the base, pinkish yellow or pinkish orange, in age orange to yellowish tan, often bruising pinkish brown, branch tips long and yellow with two to three or sometimes more short points, flesh pinkish orange, odor mild, taste

*Ramaria
formosa*

often bitter. **STALK:** Up to 6 cm tall and wide, tapered toward the base, often branched, white at the base, colored like branches above, solid, flesh becoming brittle when dry. **SPORES:** Yellowish, oblong to elliptical, roughened with warts. **HABITAT:** Single, scattered, or gregarious on ground in mycorrhizal association with hardwoods and conifers. **EDIBILITY:** Toxic, causing gastrointestinal upsets.

This coral fungus is characterized by the pinkish yellow or pinkish orange branches, yellow tips, and solid stalk. It fruits on the ground under both hardwoods and conifers. ***Ramaria rubricarnata*** is similarly colored but has smaller spores (about 11 × 4.5 μm for *R. rubricarnata* and 11 × 5.4 μm for *R. formosa*) and has amyloid stalk flesh, whereas the stalk flesh of *R. formosa* is inamyloid. ***Ramaria largentii*** has bright orange branches and tips and longer spores (about 13 × 4.5 μm).

Ramaria myceliosa

(Other name: *Phaeoclavulina curta*)

FRUITING BODY: Coral-like, 3 to 6 cm high and broad, branching profusely, the branches slender, yellowish to ochre, branch tips with two to four points and the same color as the branches, flesh whitish, odor mild, taste bitter. **STALK:** Up to 2 cm long, slender, colored like branches or whitish at the base, with attached white

Ramaria myceliosa

rhizomorphs (ropelike mycelia) binding the substrate when pulled from the ground. **SPORES:** Yellowish, elliptical, minutely spiny. **HABITAT:** Scattered, gregarious, or in arcs or rows on ground under conifers, especially common under cypress. **EDIBILITY:** Inedible.

This small yellowish coral sometimes fruits in large fairy rings under conifers. The conspicuous white rhizomorphs that bind the substrate help identify it. *Ramaria abietina* (*Phaeoclavulina abietina*) is another moderately sized ground-dwelling coral. It has yellow-brown or olive brown branches that are flattened to various degrees. Bluish green stains develop on the lower branches, especially in cold weather. *Ramaria stricta* has yellowish to pinkish tan erect, parallel, slender branches with yellow branch tips. It grows singly or in groups or rows on decaying wood (often buried). It is up to 12 cm tall and 8 cm broad. These small yellowish coral fungi are not considered for the table because of their small size and bitter taste.

Ramaria rasilispora

FRUITING BODY: Coral-like, 8 to 30 cm or more high and broad, abundantly and compactly branched, the branches pale yellow to yellowish orange, the tips short and yellow, flesh white, taste and odor mild. **STALK:** Up to 8 cm or more tall, 2 to 7 cm wide,

Ramaria rasilispora

tapering toward the base, white, solid. **SPORES:** Yellowish, cylindrical, smooth or very finely warted, inamyloid, 8 to 11.5 × 3 to 4.5 μm. **HABITAT:** Single, scattered, or in arcs and rings on the ground under hardwoods and conifers. In montane regions it fruits in spring after the snow melts; along the coast it fruits in fall and winter. **EDIBILITY:** Edible.

This large coral fungus is basically yellow with a white stalk (which is solid, not gelatinous like the bright yellow to orange *R. flavigelatinosa*). *Ramaria rasilispora* is widespread in both coastal and mountain forests. Although edible, it often carries soil and debris with it as it grows through the duff of the forest floor, or it stays partially buried, making it difficult to clean. *Ramaria magnipes* is similar, but it is a very bright yellow, has larger spores (9.5 to 13.5 × 3.5 to 5 μm), and has a massive white stalk that stains brown in age. It fruits in the mountains in spring.

Ramaria rubiginosa

FRUITING BODY: Coral-like, repeatedly branched, 6 to 12 cm high, 4 to 10 cm broad, branches white, cream colored, or pale yellow, branch tips brighter yellow, the lower branches staining red or vinaceous when bruised or in age, flesh white or pale yellow, odor and taste mild. **STALK:** Up to 4 cm wide, white at the base,

Ramaria rubiginosa

staining red or vinaceous when bruised or in age, flesh fibrous.
SPORES: Yellowish, elliptical, roughened with warts. **HABITAT:**
Single or scattered on ground in hardwood/coniferous forests.
EDIBILITY: Unknown.

The creamy or pale yellow branches and vinaceous stains are
the principal field marks of this coral fungus. ***Ramaria acrisic-***
cescens is similar, but the cream-colored branches become
browner toward the base. When immature, the branch tips of *R.
acrisiccescens* may have a pinkish or purplish tinge; in age, both
the tips and branches become brown, and it becomes difficult to
distinguish *R. acrisiccescens* from ***R. pallida,*** which is uniformly
light brown to brown, except the stalk, which is whitish at the
base. As is the case with many species in this field guide, the
name *R. pallida* is borrowed from a similar European species,
which is likely distinctly different from ours.

Ramaria sandaracina

FRUITING BODY: Coral-like, up to 14 cm high and wide, abundantly branched, branches erect and at times parallel, branches
and tips bright orange, often with a band of yellow around the
lower branches and upper stalk, the entire mushroom fading to
a dingy yellow in age, flesh yellow or orange and rubbery to

Ramaria sandaracina

brittle, odor and taste mild. **STALK:** Up to 5 cm tall and 2 cm wide, single or divided, a band of yellow above the ground, and white or pale yellow below ground level, rubbery or slightly gelatinous. **SPORES:** Yellowish, elliptical to cylindrical, roughened with warts. **HABITAT:** Scattered or gregarious on ground under conifers. **EDIBILITY:** Unknown.

   Ramaria sandaracina is generally bright orange. ***Ramaria gelatinosa*** is similar in size but has gelatinous and translucent pockets inside the stalk. The branches are pale orange or pinkish orange and the tips are yellow to pale orange. ***Ramaria flavigelatinosa*** has yellow to orange branches (the flesh of the branches is yellow in one form and salmon colored in another). The flesh of the stalk is gelatinous to cartilaginous and translucent. Other yellow or pinkish orange species, such as ***R. rasilispora*** (p. 298) and ***R. formosa*** (p. 296), have solid fibrous stalks.

Ramaria violaceibrunnea

(Other name: *Ramaria fennica* var. *violaceibrunnea*)

FRUITING BODY: Coral-like, 6 to 15 cm high, 5 to 10 cm broad, branches erect, dull olive brown, olive tan, or yellow-brown with violet tinges on the lower branches, the branch tips yellow to olive yellow, flesh white, odor mild, taste bitter. **STALK:** Up to 5

Ramaria violaceibrunnea

cm wide, often enlarged below, white with violet tinges on the upper part. **SPORES:** Yellowish, elliptical, roughened with blunt warts. **HABITAT:** Single, scattered, or in groups on ground in hardwood/coniferous forests, in mycorrhizal association with tanoaks along the coast and with conifers in the Pacific Northwest. **EDIBILITY:** Unknown.

The pale brown ground color of the branches and violet tinges of the lower branches and upper stalk are distinctive. The branches are generally erect, giving this coral an upright appearance. ***Ramaria pallida*** is uniformly light brown to brown except for the very base of the stalk, which is whitish. It grows singly or scattered on ground in hardwood/coniferous forests. Like many mushrooms, the name *R. pallida* is borrowed from a similar European species. Our local species, although clearly related to *R. pallida*, is likely a distinct and as yet unnamed species.

Tremellodendropsis tuberosa
FRUITING BODY: Coral-like fingers 3 to 6 cm tall, up to 5 cm broad, unbranched or sparsely branched, the ends of the arms rounded and often flattened, upright, white becoming brownish in age, tough. **STALK:** White, tough, the base covered with white

Tremellodendropsis tuberosa

mycelium, flesh white, not changing color when bruised. **SPORES**: White, spindle shaped, smooth, inamyloid. **HABITAT**: Scattered to gregarious on ground under Coast Redwood, oaks, cypress, and other trees in coastal forests. **EDIBILITY**: Unknown.

The flattened branches rising from a tough base and growth on the ground characterize this small coral fungus. The branch tips are rounded or blunt, unlike the finely toothed branch tips of **Clavulina cristata,** which is also uniformly white, at least when young. **Clavulina rugosa** has wrinkled white branches that end in blunt tips, or at least lack the toothed tips of *C. cristata*. It is typically sparsely branched, but the branches are not flattened as in *T. tuberosa*. **Ramariopsis kunzei** (p. 293) is a profusely branched white coral fungus; the forks of the branches are usually U-shaped, and the branches end in blunt or pointed tips. It grows on the ground in mixed woods and is especially common under conifers.

Jelly Fungi

The jelly fungi are a loose collection of species known to have evolved from several different ancestors. As a group, they are characterized by their gelatinous or rubbery texture when moist.

Some are shapeless and convoluted, whereas others are erect or gumdroplike. The fruiting bodies dry to a papery crust in dry weather and quickly rehydrate in rain. None of the jelly fungi are known to be poisonous, but most are too small for the table. However, a few larger species are collected or commercially grown. The Wood Ear, for example, is one of the most popularly cultivated mushrooms in the world. Although flavorless on its own, it soaks up the flavor of any dish and provides interesting texture. Some jelly fungi decay wood; others are parasites of other fungi that decay wood.

a. Fruiting body light colored
    b. Gelatinous and convoluted when moist, yellow .... **Tremella aurantia**
    b. Gelatinous and gumdroplike, yellow ....**Guepiniopsis alpina**
    b. Rubbery and more-or-less erect
        c. Yellow to orange, erect, clustered on wood ....**Calocera spp.** (see *Tremella aurantia*)
        c. Translucent white, underside of cap minutely spiny ....**Pseudohydnum gelatinosum**

a. Fruiting body dark colored
    d. Rubbery, thin, brown, earlike............... **Auricularia auricula**

Calocera cornea

d. Gelatinous
  e. Black, lobed, fusing with others to form masses..............
  .... **Exidia glandulosa**
  e. Brown, wavy, seaweedlike ....
  ....**Tremella foliacea** (see *Tremella aurantia*)

Auricularia auricula

Wood Ear

FRUITING BODY: 2 to 15 cm broad, sometimes fused into masses of indefinite size, usually hanging downward from a point of attachment, cup or ear shaped, tan to reddish brown, becoming nearly black when dry, upper sterile surface with a downy covering and often veined or ridged, inner fertile surface smooth or wrinkled with intersecting ridges, texture rubbery-gelatinous when moist and expanded, shrinking and becoming thin and brittle when dry. **SPORES:** White, cylindrical to sausage shaped, smooth. **HABITAT:** Single, gregarious, or clustered on dead wood. **EDIBILITY:** Edible and commercially grown. It is rather flavorless by itself, but it quickly acquires the taste of any seasoning and provides an interesting consistency to dishes.

Auricularia auricula

The brown earlike fruiting body and growth on wood are the primary features of this cosmopolitan jelly fungus. It shrivels and shrinks when dry and expands greatly when moistened to release the spores. The Wood Ear could be mistaken for a cup fungus, but the concave fertile portion of cup fungi faces up, whereas *A. auricula* usually hangs with the concave, cuplike fertile portion pointing downward. Similar-appearing jelly fungi include **Tremella foliacea,** which has wavy, seaweedlike brown lobes and spherical spores. **Guepinia helvelloides** (*Phlogiotis helvelloides*) is pinkish orange and funnel-like or tonguelike. It grows on the ground or rotted wood. Also see **Exidia recisa** under **E. glandulosa** (below).

Exidia glandulosa
Black Witch's Butter

FRUITING BODY: 1 to 2 cm broad but often becoming laterally fused to form irregular masses up to 20 cm or more long, each fruiting body initially cushion shaped, becoming wrinkled, convoluted, and brainlike, often with multiple lobes, pale brown at first becoming reddish brown and finally black at maturity, the spore-bearing surface smooth to roughened with minute warts, the texture rubbery-gelatinous when moist but shrinking and brittle and crusty when dry, odor and taste mild. **STALK:** Absent. **SPORES:** Colorless, smooth, cylindrical to curved and sausage shaped, produced on elliptical, septate basidia. **HABITAT:** Single or gregarious, often in rows on rotting hardwood logs and branches. **EDIBILITY:** Unknown.

This black jelly fungus is an excellent example of fungal adaptation to arid climates. It may remain shriveled and dormant for long periods of drought, but within hours of a rain the gelatinous fruiting body swells and begins to sporulate, thus assuring that the spores will have moisture to germinate and colonize new substrates. Under dry conditions the fruiting body once again shrinks and dries to an almost unrecognizable film as it awaits the next rain. *Exidia recisa* is reddish brown in color, cone or ear shaped, and hangs from fallen branches. It has a veined rather than warted spore-bearing surface. *Exidia nucleata* is smaller, translucent-white or white tinged with pink, and has white, nodulelike structures in the center of the fruiting body.

Exidia glandulosa

Guepiniopsis alpina
(Other name: *Heterotextus alpinus*)
FRUITING BODY: Up to 1.5 cm in diameter, cone shaped, hanging from a narrow attachment, appearing like a flabby gumdrop, bright yellow to orange, smooth, viscid, and gelatinous, drying

Guepiniopsis alpina

reddish orange and rigid, flesh gelatinous. **STALK:** A mere pointed attachment. **SPORES:** Pale yellow, sausage shaped, septate, smooth. **HABITAT:** Scattered, gregarious, in rows, or in clusters on barkless conifer wood, especially common at high elevations in spring soon after the snow melts but also found in other habitats. **EDIBILITY:** Unknown but too small to be of value.

The bright yellow color of the gumdroplike fruiting body that hangs from a pointlike attachment quickly identifies this jelly fungus. It is very common at high elevations in spring, often found attached to the wood of fallen conifers. Other yellow cushionlike jellies, such as *Dacrymyces capitatus* and *D. stillatus,* are smaller and are not pendulant on their substrate (see p. 309). *Bisporella citrina* (p. 402) is small (less than 3 mm across), yellow, and disk shaped and fruits in swarms on dead barkless hardwood. Unlike *H. alpinus*, which produces spores on basidia, *B. citrina* produces spores in asci (sacs).

Pseudohydnum gelatinosum

Toothed Jelly Fungus

FRUITING BODY: Tongue, spatula, or fan shaped, erect and stalked or shelflike from the substrate, cap 1 to 4 cm broad, translucent-white, faintly browning in age, the upper surface smooth to minutely velvety, the fertile underside covered with tiny down-

Pseudohydnum gelatinosum

ward-pointed spines or teeth, similar to the roughness of a cat's tongue, the texture flabby-flexible to rubbery-gelatinous. **STALK:** Usually present, up to 4 cm long, lateral, tapering toward the base, colored like the cap, rubbery-gelatinous. **SPORES:** Colorless, subglobose to globose, smooth. **HABITAT:** Single, scattered, or gregarious on rotted logs, branches, and litter under conifers. **EDIBILITY:** Edible and tasteless but can be candied and served as a dessert.

This peculiar little fungus has a unique combination of characteristics: a translucent-white color, texture of rubber, and small teeth on the underside similar to a cat's tongue. Although it resembles a tooth fungus, the divided basidia clearly place it with the jelly fungi. Other fungi with teeth such as ***Auriscalpium vulgare*** (p. 279) and ***Hydnum repandum*** (p. 283) lack the gelatinous texture, have simple basidia, and are differently colored. The name *Pseudohydnum*, indicating a *Hydnum* look-alike, is appropriate. Other jelly fungi are rubbery, but none have small toothlike projections on the undersurface.

Tremella aurantia
Witch's Butter

FRUITING BODY: Up to 10 cm broad, at first cushion shaped, becoming wrinkled, brainlike and finally convoluted and lobed, when moist, bright yellow to golden yellow, viscid, and gelatinous, drying reddish orange and rigid. **STALK:** Absent. **SPORES:** Pale yellow, oval to nearly round, smooth. **HABITAT:** Single, scattered, or clustered on hardwood, parasitizing the wood rotter *Stereum hirsutum*. **EDIBILITY:** Edible but flavorless; it can be used in soups or hot dishes where texture is desired.

The flabby, yellow, amorphous fruiting body of this jelly fungus is a common sight in western forests after rains. It seems to appear out of nowhere after rains cause it to swell and sporulate. It is frequently confused with the similar ***Dacrymyces palmatus***, which grows as a saprobe on dead, barkless conifer wood. That fungus has Y-shaped basidia, whereas *T. aurantia* has longitudinally septate basidia. ***Tremella foliacea*** is similar but brown. Other yellowish to orange jelly fungi include ***Dacrymyces capitatus***, which produces tiny (up to 2 mm broad) cushion-shaped, rooted fruiting bodies on conifers and hardwoods; ***D. stillatus***, a conifer wood inhabitant with orange to brownish, sessile,

Tremella aurantia

cushion-shaped fruiting bodies up to 3 mm broad; ***Calocera cornea*** (p. 304), a small jelly fungus with erect cylindrical fruiting bodies 5 to 15 mm high; and ***C. viscosa,*** a branched species up to 7 cm high. All of these grow on wood.

Thelephora terrestris
Earth Fan

FRUITING BODY: Vaselike, fanlike, or irregularly shaped caps 2 to 5 cm broad, often produced in compound clusters, arising horizontally or vertically from a base, reddish brown to brown, darkening in age, the surface silky-fibrillose, the actively growing edge often fringed and paler than the rest of the cap, tough, thin, fibrous, and dry, odor mild. **SPORE-BEARING SURFACE:** Smooth or wrinkled, poreless, brown. **STALK:** A common base supporting clusters or individual caps, tough, brown, and short. **SPORES:** Purplish brown, irregularly shaped, warted. **HABITAT:** Single or more typically in clusters on the ground in woods, along paths, and in landscaped areas; sometimes climbing stems of plants. **EDIBILITY:** Unknown but too tough and leathery to be of value.

This easily overlooked fungus sometimes forms large compound clusters of fan- or vase-shaped caps, resulting in an ir-

Thelephora terrestris

regularly shaped rosettes; other times, it forms a single vase on the ground or perched on wood or the stem of a plant. ***Thelephora palmata*** is characterized by tough, flattened, wrinkled, purplish brown branches with flat, palmlike tips. Clusters of branches arise from a common base. The actively growing margin is whitish or paler than the tough and fibrous branches. It has an odor of garlic. All *Thelephora* species are thought to be mycorrhizal.

Stereum hirsutum
False Turkey Tail

CAP: 1 to 4 cm broad or often appearing larger when fused with others, shape highly variable—irregular, fanlike, semicircular, resupinate, or effused-reflexed; thin, tough, wavy, hairy overall when young but the hairs falling off in bands, exposing concentric zones of orange, orange-brown, and yellow-brown, in age becoming duller and gray. **SPORE-BEARING SURFACE:** Bright orange to dull orange-brown, smooth, sometimes zoned. **STALK:** Absent. **SPORES:** White, cylindrical, smooth, amyloid. **HABITAT:** In rows, tiers, and clusters, often fused together on hardwood branches and logs causing a white rot. **EDIBILITY:** Inedible.

Stereum hirsutum

The thin, tough, wavy, brownish orange cap that has zones of various colors is found on dead hardwood. The orange underside (fertile surface) is smooth, lacking gills, pores, and spines. With age the bright orange colors eventually fade to gray and the fungus resembles ***Trametes versicolor***, the Turkey Tail (p. 356). That fungus, however, has minute pores on the fertile surface. Similar species with a smooth spore-bearing surface include ***Stereum ostrea***, which has a red- and brown-zoned cap up to 7 cm broad that usually grows individually, and **S. ochraceoflavum**, which has a buff to pale tan, faintly zoned cap covered uniformly with long white hairs. The fertile surface is pale tan and may be faintly zoned. It also grows on dead hardwood but favors sticks and small branches.

Phlebia radiata
Wrinkled Crust

FRUITING BODY: Resupinate (flattened against its substrate), orange or pinkish cinnamon, conspicuously wrinkled, waxy, circular to irregular, up to 10 cm in diameter, attached firmly to the substrate. **SPORES:** White, sausage shaped, smooth, inamyloid. **HABITAT:** Single to clustered, saprobic on decaying wood of

Phlebia radiata

conifer and hardwood trees, causing a white rot. **EDIBILITY:** Inedible.

This common fungus represents a very large group of resupinate crust fungi that are extremely important in recycling wood and litter. *Botryobasidium vagum* is a white to brown crust fungus composed of loosely woven, thick hyphae that branch at right angles from one another. *Serpula lacrymans* and *Meruliporia incrassata* are the notorious dry rot fungi responsible for destroying indoor construction wood, even in seemingly dry conditions. To find needed moisture, these fungi extend long rhizomorphs to areas such as damp ground. Both cause a brown rot of wood by preferentially breaking down cellulose, leaving the modified lignin intact. In contrast, other crust fungi are mycorrhizal with trees. *Piloderma bicolor,* for example, is mycorrhizal with pines. It fruits on the underside of logs and has yellow conspicuous rhizomorphs that penetrate the soil.

Leccinum manzanitae
Manzanita Bolete

CAP: 7 to 20 cm in diameter, rounded to convex and finally plane, red to reddish brown, smooth or faintly fibrillose, surface viscid when moist, flesh white, often bruising gray or bluish gray, odor and taste mild. **PORES:** White becoming pale olive, bruising brown. **STALK:** 8 to 18 cm long, 2 to 4 cm wide, equal to clavate, firm, white and covered with black tufts of hairs (scabers), often staining bluish where bruised. **SPORES:** Brown, elliptical to spindle shaped, smooth. **HABITAT:** Single or scattered on the ground near manzanita and madrone. **EDIBILITY:** Edible but of poor quality.

The reddish cap, dark scabers on the whitish stalk, and mycorrhizal habitat with manzanita, madrone, and possibly other tree species are the field marks of this robust species. **Leccinum insigne** (p. 424), the Aspen Bolete, has an orange to brownish orange, smooth or minutely pubescent cap. The pores are white to olive yellow, and the white or off-white stalk is covered with brown to black scabers. Like *L. manzanitae*, the stalk often stains bluish when bruised. *Leccinum insigne* fruits on the ground in summer or fall in mycorrhizal association with aspens. It is edible. The brown-capped **L. scabrum** is restricted to ornamental birch in cultivated landscapes.

Leccinum manzanitae

Porphyrellus porphyrosporus

Porphyrellus porphyrosporus
(Other names: *Tylopilus porphyrosporus, T. pseudoscaber*)
CAP: 6 to 15 cm in diameter, convex to broadly convex, uniformly dark brown, dull, dry, smooth or velvety, flesh whitish, often but not always turning grayish pink to blue when bruised. **PORES:** Dark brown, round or angular, up to 1 mm in diameter and occasionally larger, often staining blue and then brown when bruised. **STALK:** 7 to 16 cm long, 1.5 to 3 cm wide, equal or with an enlarged base, firm, dry, mostly colored like the cap but the base whitish, flesh whitish, bruising pink, brown, or blue, partial veil absent. **SPORES:** Reddish brown, elliptical to spindle shaped, smooth. **HABITAT:** Single or scattered under conifers. **EDIBILITY:** Unknown.

The dark brown cap, pores, and stalk, reddish brown spores, and bluing reaction when bruised are distinctive features of this species. Although the bluing reaction is variable from specimen to specimen, when *P. porphyrosporus* is wrapped in wax paper, the paper is usually stained blue. *Tylopilus indecisus* is a widespread species that has a brown cap, pink pore surface that bruises brown, pinkish spores, and a more-or-less equal stalk that is finely reticulate near the apex. It grows on the ground with hardwoods, especially oaks.

Chalciporus piperatus

Chalciporus piperatus

(Other name: *Boletus piperatus*) Peppery Bolete

CAP: 2 to 7 cm in diameter, convex to broadly convex to plane, orange-brown to reddish brown, surface moist but soon dry, smooth to minutely fibrillose, flesh yellowish, darkening to brownish when bruised but not becoming blue, taste peppery. **PORES:** Angular, up to 2 mm in diameter, sometimes slightly running down the stalk, orange-brown, darkening when bruised. **STALK:** 2 to 6 cm long, about 0.5 to 1 cm wide, colored like the cap, equal or tapering downward, flesh bright yellow near the base, yellow mycelium anchoring the fruiting body to the ground. **SPORES:** Brown, spindle shaped, smooth. **HABITAT:** Single or scattered in small groups under pines. **EDIBILITY:** Unknown, but the peppery taste may be a deterrent.

The distinctive features of this bolete are its small size, overall orange-brown color, mass of yellow mycelium adhering to the base of the stalk, and peppery taste. It and the similar ***Chalciporus piperatoides*** (*Boletus piperatoides*) are the smallest boletes in the West. The distinguishing feature of the latter is the bluing of the pores when bruised. Although the peppery taste might be used as a spice, the edibility of these mushrooms has not been firmly established.

Boletus Species

Boletes are a conspicuous part of the fungal flora of western forests. They are typically robust and carry themselves with a stately appearance. As a group these species are united by a spongelike layer of tubes attached to the underside of the caps, nonviscid texture (with rare exceptions), and olive brown spores. The layer of tubes is easily separated from the cap, which distinguishes the boletes from the polypores, which have a layer of tubes inseparable from the cap. Boletes have no ring on the stalk, which is equal to bulbous. The King Bolete or Porcini (*Boletus edulis*), a choice edible, is one of the most sought-after mushrooms in western forests. Species of *Suillus* differ from *Boletus* by some combination of a viscid cap, stalk with a ring, pores arranged radially, and minute paintlike dots on the stalk, features generally absent in *Boletus*. Similar genera include *Leccinum*, which has tufts of blackish hairs on the stalk, *Porphyrellus*, which has pinkish to reddish brown spores, and *Chalciporus*, with yellowish flesh and a peppery taste. The boletes and relatives are mycorrhizal with trees.

a. Stalk strongly reticulate
    b. Pores bruising blue
        c. Pores red .... **Boletus eastwoodiae**
        c. Pores not red
            d. Cap yellowish.................................... **B. appendiculatus**
            d. Cap rose colored or pinkish...
            ....**B. regius** (see *B. appendiculatus*)
    b. Pores not bruising blue
        e. Cap brownish, with a white bloom when young...............
        ....**B. regineus**
        e. Cap brownish, without a white bloom when young.........
        .... **B. edulis**

a. Stalk weakly or nonreticulate
    f. Pores reddish
        g. Pores orange-red; cap reddish brown ......**B. amygdalinus**
        g. Pores dark red; cap dark brown...
        ....**B. erythropus** (see *B. amygdalinus*)
    f. Pores yellowish

h. Cap viscid; pores not bluing when bruised......................
.... *B. flaviporus*
h. Cap not viscid
  i. Pores bluing
    j. Cap tan......................................*B. rubripes*
    j. Cap blackish ....*B. zelleri*
    j. Cap brownish ....*B. chrysenteron*
  i. Pores not bluing or bluing erratically
    k. On wood; pores not bluing when bruised................
    .... *B. mirabilis* (see *B. chrysenteron*)
    k. On ground; pores sometimes bluing when bruised
    ....*B. subtomentosus*

Boletus amygdalinus

CAP: 4 to 12 cm in diameter, convex, reddish brown to red, dry, smooth, the margin often wavy, flesh yellow, bruising blue. **PORES:** Red when young becoming reddish orange, bruising blue. **STALK:** 4 to 8 cm long, 2 to 3 cm or more wide, equal or slightly thicker at the base, a mixture of red and yellow, smooth (not reticulate), dry, flesh yellow, bruising blue. **SPORES:** Olive brown, elliptical, smooth. **HABITAT:** Single or scattered under

Boletus amygdalinus

oaks and other hardwoods; particularly common in southern California oak woodlands. **EDIBILITY:** Possibly poisonous.

The reddish orange pores, dry reddish brown cap, red and yellow nonreticulate stalk, and bluing reaction of all parts of the fruiting body identify this mushroom. Like most red-pored boletes, it should be considered poisonous. *Boletus erythropus* (p. 425) is a similar species that grows in mixed woods in the West. It has a dark brown or reddish brown cap, yellow and red nonreticulate stalk, and red pores. All parts of the fruiting body bruise blue extremely rapidly. *Boletus frostii* has a red cap, red pores that bruise blue, and a red reticulate stalk. It grows in the mountains of Arizona, in the Rocky Mountains, and farther east.

Boletus appendiculatus

CAP: 6 to 20 cm in diameter, convex to broadly convex, yellow, yellowish tan, or brown, often with reddish brown or rusty stains in age, surface dry or slightly viscid when moist, smooth or finely pubescent when young, flesh firm and yellowish, slowly bruising blue or not at all, odor and taste mild. **PORES:** Lemon yellow, duller in age, usually bruising blue. **STALK:** 5 to 10 cm

Boletus appendiculatus

long, 3 to 6 cm wide, usually bulbous, sometimes with a tapered base, yellow, sometimes with red or brownish stains, solid and very firm, the upper part of the stalk reticulate. **SPORES:** Olive brown, spindle shaped, smooth. **HABITAT:** Single or scattered under hardwood, especially oaks. **EDIBILITY:** Edible and choice.

This mushroom is identified by its large size, yellowish tan cap, yellow pores that usually bruise blue, and reticulate stalk. *Boletus regius* is similar in both size and stature but has a rose or deep pink-colored cap. It also has a yellow reticulate stalk and yellow pores that usually turn blue when bruised. Both species are good edibles. *Boletus edulis* (p. 322) also has a reticulate bulbous stalk, but it has a brownish cap and pale pores that never bruise blue.

Boletus chrysenteron

(Other name: *Xerocomus chrysenteron*)

CAP: 4 to 10 cm in diameter, convex to flat, olive brown to brown, surface dry and velvety when young, usually areolate (surface divided by cracks), the valleys of the cracks pink, especially near the margin, flesh sometimes bluing slowly when bruised. **PORES:** Yellow to olive brown, relatively large (1 to 2 mm in diameter), usually bluing slowly when bruised. **STALK:** 5 to 10 cm long, 1 to 2 cm wide, equal, smooth or longitudinally striate, yellow at the apex and reddish below. **SPORES:** Olive brown, elliptical to spindle shaped, smooth. **HABITAT:** Single or scattered in hardwood/coniferous forests. **EDIBILITY:** Edible.

The areolate surface of the brownish cap with pink tints in the cracks, the red and yellow stalk, and the bluing of the pores identify this common bolete. *Boletus zelleri* is similar but has a black or very dark olive brown cap that may develop red tones in age (p. 327). The surface of the cap is often somewhat wrinkled and does not crack much if at all. The stalk is yellow and red or red throughout. Like the pores of *B. chrysenteron*, the pores of *B. zelleri* usually turn blue when bruised. *Boletus mirabilis*, which fruits on rotting wood, has a maroon-brown cap and stalk and yellow pore surface that does not blue when bruised. *Boletus dryophilus* has a red cap, yellow pores that bruise blue, a stalk that is yellow at the apex and red below, and an oak habitat. All of these mushrooms are edible.

Boletus chrysenteron

Boletus eastwoodiae

(Other name: *Boletus satanas*) Satan's Bolete

CAP: 10 to 25 cm in diameter, convex, olive buff or olive tan, developing pink tones in age, dry, the margin curved inwardly in young specimens and expanding in age, flesh yellow to buff, bruising blue, odor and taste mild. **PORES:** Red becoming orange in age, bruising blue. **STALK:** 7 to 15 cm long, 3 to 6 cm wide at apex, with a massive, more-or-less rounded basal bulb up to 15 cm or more wide, upper portion reticulate and pink, lower portion paler pink or buff, fading in age, solid, bruising blue. **SPORES:** Olive brown, elliptical, smooth. **HABITAT:** Single or scattered under oaks in coastal and montane forests. **EDIBILITY:** Not recommended, possibly poisonous.

Satan's Bolete is easily identified by the reddish pores, olive buff cap, bluing reaction of all parts when bruised, and especially the large, bulbous, reticulate stalk. *Boletus pulcherrimus* also has red pores and a brown or olive brown cap with pinkish tones and a conspicuously reticulate stalk. The base of the stalk, however, is swollen, not bulbous. Like *B. eastwoodiae*, all parts of the fruiting body bruise blue. For other boletes with red pores, see *B. amygdalinus* (p. 318).

Boletus eastwoodiae

Boletus edulis **(p. vi)**

King Bolete, Porcini, Cep

CAP: 6 to 25 cm in diameter, broadly convex, reddish brown, chestnut brown, or yellow-brown, smooth, not bluing when bruised. **PORES:** Small, whitish at first, becoming yellow or olive

Boletus edulis

yellow in age, not bluing when bruised. **STALK:** 8 to 20 cm long, 3 to 10 cm wide, bulbous becoming equal in age, off-white to pale tan, the upper part reticulate (netted). **SPORES:** Olive brown, elliptical to spindle shaped, smooth. **HABITAT:** Single or scattered in hardwood/coniferous forests, especially common under pines in both coastal and montane forests. **EDIBILITY:** Edible and choice.

The King Bolete is one of the most highly prized mushrooms in temperate forests around the world. It is identified by its reddish brown cap, reticulate stalk, and whitish to olive yellow pores that do not blue when bruised. It is often large and robust. In young specimens, the base of the stalk is typically quite thick. Two look-alikes are equally edible: **B. rex-veris**, formerly known as *B. pinophilus*, is found in the mountains in spring; and **B. barrowsii** is very similar to *B. edulis*, but it is whitish in color. It is most common under pines after warm, summer monsoon rains in the mountains of the Southwest but may be encountered in fall in Sierra Nevada and occasionally in forests of coastal California.

Boletus flaviporus

CAP: 6 to 15 cm in diameter, convex to broadly convex to plane, reddish brown, surface viscid (in dry weather look for adhering debris), smooth but streaky, not bluing when bruised. **PORES:** Bright lemon yellow, not turning color when bruised. **STALK:** 6 to 15 cm long, 1 to 2 cm wide, equal or tapering downward, smooth, viscid when moist, the apex yellow but otherwise a mixture of yellow and reddish brown, sometimes slightly reticulate from tubes running down the stalk, white mycelia often visible at the base of the stalk. **SPORES:** Olive brown, elliptical to spindle shaped, smooth. **HABITAT:** Single or scattered in mixed woods, especially under oak. **EDIBILITY:** Unknown.

The viscid reddish brown cap and bright yellow pores are distinctive. The viscid cap may cause confusion with *Suillus*, since the caps of species of *Boletus*, as a general rule, are dry and those of *Suillus* are usually viscid. Specifically, *B. flaviporus* might be confused with **Suillus brevipes** (p. 329), but that species has dull yellow pores, a dark brown cap, and a relatively short stalk. In addition, *S. brevipes* is associated with pines rather than oaks, like *B. flaviporus*. **Boletus citriniporus** also has bright yellow pores, but it has a dry, dark brown cap.

Boletus flaviporus

Boletus regineus
(Other name: *Boletus aereus*) Queen Bolete
CAP: 5 to 18 cm in diameter, broadly convex becoming nearly flat in age, the margin inrolled at first, when young the surface brown to chestnut brown and covered, at least in part, with a whitish bloom (appearing like a dusting of fine white powder), in age losing the bloom and becoming cinnamon to reddish brown, flesh white, not bluing when bruised, odor and taste mild. **PORES:** Small, whitish at first, becoming yellow or olive yellow in age, not bluing when bruised. **STALK:** 5 to 15 cm long, 3 to 6 cm wide, clavate or equal in age, off-white to pale tan, the upper part reticulate (surface netlike), otherwise smooth, partial veil absent. **SPORES:** Olive brown, elliptical to spindle shaped, smooth. **HABITAT:** Single, scattered, or gregarious in mixed hardwood/coniferous forests. **EDIBILITY:** Edible and choice.

The Queen Bolete is recognized by the cinnamon brown to reddish brown cap, reticulate stalk, whitish to olive yellow pores that do not blue when bruised, and especially the white bloom on the young caps. In age it looks like **B. edulis,** the King Bolete (p. 322), but the latter favors pines, whereas *B. regineus* occurs in mixed woods. A similar species, **B. rex-veris,** is found in the

Boletus regineus

mountains in spring. It sometimes fruits in clusters, unlike *B. regineus* and *B. edulis.*

Boletus rubripes

CAP: 6 to 20 cm in diameter, broadly convex, tan, dry, dull, and velvety, sometimes cracked in age, flesh white to pale yellow, bluing rapidly when bruised, odor mild, taste bitter. **PORES:** Yellow, oval or angular, bluing when bruised. **STALK:** 6 to 15 cm long, 2 to 5 cm wide, equal or with a swollen base, apex yellow, reddish below, becoming all red in age, not reticulate but may be longitudinally striate in age. **SPORES:** Olive brown, elliptical to spindle shaped, smooth. **HABITAT:** Single or scattered in coastal and montane hardwood/coniferous forests. **EDIBILITY:** Inedible due to the bitter taste.

This robust and beautiful bolete has a partially or predominately red stalk, yellow pores that bruise blue, a dry and dull tan cap, and bitter taste. **Boletus calopus,** another large bolete with a red and yellow stalk and dry, dull tan cap, differs by having a reticulate stalk. It has pale yellow pores and a distinctly and quickly bitter taste. It grows under conifers at high elevations in spring or summer after rains. **Boletus edulis** (p. 322) also has a reticulate stalk, but its pores never turn blue, and it does not have a bitter taste.

Boletus rubripes

Boletus subtomentosus

CAP: 5 to 15 cm in diameter, convex to plane, dull brown, the surface dry and velvetlike from minute hairs, sometimes with cracks in age that are buff or rarely pink, flesh unchanging or bluing weakly when bruised. **PORES:** Bright yellow when young,

Boletus subtomentosus

becoming dull or dirty yellow, relatively large (up to 2 mm or more wide), sometimes running down the stalk a short distance, bluing weakly or not at all when bruised. **STALK:** 4 to 12 cm long, 1 to 2 cm wide, equal or narrowing toward the base, smooth but the apex ridged from adhering pores, firm, yellow or buff, often with brown stains, the base with adhering yellow mycelium. **SPORES:** Olive brown, elliptical to spindle shaped, smooth. **HABITAT:** Single or scattered in mixed hardwood/coniferous forests. **EDIBILITY:** Edible but of poor quality.

The dull brown velvety cap, yellowish pores that may or may not stain blue, and absence of red colors on the stalk are the primary features of this bolete. The cap does not crack as often as *B. chrysenteron* (p. 320), and the stalk is neither bulbous nor reticulate like *B. edulis* (p. 322). The bluing reaction is not a reliable characteristic because some specimens do not bruise blue at all. *Boletus citriniporus* is a small bolete with a dark brown cap about 6 cm broad and bright yellow pores that never turn blue.

Boletus zelleri

(Other name: *Xerocomus zelleri*)

CAP: 4 to 12 cm in diameter, convex to nearly flat, dark olive brown, dark brown, to almost black, sometimes developing dark maroon tints in age, surface dry and velvety when young, smooth or wrinkled, occasionally cracked in age, flesh pale yellow, often bluing slowly when bruised, odor and taste mild. **PORES:** Yellow to olive yellow, relatively large (1 to 2 mm in diameter), usually bluing slowly when bruised. **STALK:** 5 to 12 cm long, 1 to 3 cm wide, equal, dry, smooth or longitudinally striate, red or red on a yellow background, the base often yellow. **SPORES:** Olive brown, spindle shaped, smooth. **HABITAT:** Single or scattered on ground in mixed hardwood/coniferous forests, especially under conifers. **EDIBILITY:** Edible.

Boletus zelleri has a black or very dark olive brown cap that may develop red tones in age. The surface of the cap is often somewhat wrinkled and does not crack much if at all. The stalk is mostly red or red and yellow. *Boletus chrysenteron* (p. 320) is similar, but the brownish cap cracks in age, exposing pink tints of the cuticle. It has a red and yellow stalk, and like the pores of *B. zelleri*, the pores of *B. chrysenteron* sometimes, but not al-

Boletus zelleri

ways, turn blue when bruised. **Boletus mirabilis** has a maroon-brown cap and stalk and yellow pore surface that does not blue when bruised. It fruits on wood.

Suillus Species

These are the slippery jacks, so named because of the slimy surface of the caps of some species. Like *Boletus*, they have a spongelike layer of tubes attached to the underside of the caps. *Suillus* species have a unique combination of characteristics, not all of which are present in all species: viscid cap, stalk with a ring, pores often arranged radially, and minute paintlike dots on the stalk. All are mycorrhizal with conifers.

a. Stalk glandular dots absent
   b. Veil absent...**Suillus brevipes**
   b. Veil usually present
      c. Cap viscid; sparsely fibrillose or smooth............................
      ....**S. caerulescens**
      c. Cap usually dry; covered with red fibrils .... **S. lakei**

a. Stalk glandular dots present
   d. Cap smooth

Suillus fuscotomentosus

> e. Cap white, gray, or brownish .... **S. pungens**
> e. Cap orange-brown to cinnamon brown..... **S. granulatus**
> d. Cap fibrillose
>> f. Flesh bluing when bruised .... **S. tomentosus**
>> f. Flesh not bluing when bruised ....
>> ....**S. fuscotomentosus** (see *S. tomentosus*)

Suillus brevipes

CAP: 4 to 10 cm in diameter, convex, dark brown, fading in age, surface smooth, viscid, often shiny, not bluing when bruised. **PORES:** Yellow when young, aging olive yellow, one to two pores per millimeter, not bruising blue. **STALK:** 1 to 6 cm long, 1 to 3 cm wide, equal or slightly clavate, white to yellow, smooth, glandular dots usually not visible, not bluing when bruised, partial veil absent. **SPORES:** Brown, elliptical, smooth. **HABITAT:** Scattered or gregarious under pines. **EDIBILITY:** Edible.

The viscid to slimy dark brown cap, yellow pores, and relatively short smooth stalk are the hallmarks of this slippery jack. It is often abundant under pines in the coastal forests along the northern California coast in fall and early winter. This species might be confused with **Boletus flaviporus** (p. 323), which has a reddish brown viscid cap, a relatively long stalk, and a bright

Suillus brevipes

yellow pore surface, whereas the pore surface of *S. brevipes* is a dull or pale yellow to olive yellow in age. The color of the stalk of *B. flaviporus* is usually flushed with reddish brown, whereas the stalk of *S. brevipes* is uniformly whitish or yellow. Finally, *B. flaviporus* is mycorrhizal with oaks, whereas *S. brevipes* is associated with pines.

Suillus caerulescens

CAP: 5 to 15 cm in diameter, convex to plane, yellow to pale yellow-orange with scattered reddish brown fibrils, viscid when moist, flesh yellowish, not bluing when bruised. **PORES:** Yellow, aging or bruising yellowish brown, browning where bruised, pores 1 mm or more in diameter and often radiating from the stalk, not bluing when bruised. **STALK:** 2 to 8 cm long, 1 to 3 cm wide, yellow with reddish streaks and spots below the ring; the ring yellowish, fibrillose, and often poorly defined; the bottom part of the stalk slowly bruising blue (although sometimes not bluing in old specimens). **SPORES:** Cinnamon brown, elliptical, smooth. **HABITAT:** Scattered to gregarious on ground under Douglas-fir. **EDIBILITY:** Edible but of poor quality.

The yellowish cap with scattered fibrils, presence of a partial veil, and development of blue stains in the base of the stalk are

Suillus caerulescens

the important characteristics of *S. caerulescens*. **Suillus lakei** (p. 332) is similar in many respects, but it has abundant reddish fibrils on a yellow background, resulting in a striking color combination. Like *S. caerulescens*, it has pores that stain brown, a partial veil, and a lower stalk that stains blue when cut. Both of these mushrooms fruit under Douglas-fir in fall and winter, and both are edible but of poor quality.

Suillus granulatus

CAP: 4 to 12 cm in diameter, convex to nearly flat, orange-brown to cinnamon brown, often becoming spotted or mottled in age as the surface color breaks up, viscid when moist, shiny when dry, smooth, flesh white to pale yellow, not bluing when bruised, odor and taste mild. **PORES:** White when young but soon yellowish, sometimes exuding droplets of liquid, not bluing when bruised. **STALK:** 4 to 8 cm long, 1 to 2 cm thick, equal or tapered at the base, firm, solid, white becoming yellow, sometimes stained red, covered with red or brown glandular dots, not bluing when bruised, partial veil absent. **SPORES:** Cinnamon, elliptical to spindle shaped, smooth. **HABITAT:** Single, scattered, or gregarious under pines, widespread. **EDIBILITY:** Edible but of poor quality.

Suillus granulatus

The orange-brown cap, presence of glandular dots on the stalk, and the absence of a veil are the principal field marks of this common slippery jack that grows in mycorrhizal association with a variety of pines. It might be confused with **S. pungens** (p. 333), but that species has a grayish cap with olive tones when young and a more limited distribution. It commonly occurs with Monterey pines.

Suillus lakei

CAP: 5 to 15 cm in diameter, convex to nearly flat, the margin inrolled when young and sometimes edged with partial veil remnants, surface covered with red to reddish brown fibrils on a yellow to yellow-orange background, typically dry but sometimes viscid in wet weather, the cuticle peeling easily, flesh yellowish, sometimes bruising pinkish but not bluing. **PORES:** Yellow to ochre, angular, some up to 3 mm wide and radially arranged, slightly decurrent, bruising yellowish brown to brown. **STALK:** 3 to 8 cm long, 1 to 3 cm wide, equal, yellow with reddish stains and fibrils below the ring, glandular dots absent; the ring thin, white to yellowish, ragged and often poorly defined; the bottom part of the stalk slowly bruising blue (although sometimes not bluing in old specimens). **SPORES:** Brown, elliptical to spindle shaped, smooth. **HABITAT:** Scattered to gregarious on ground under Douglas-fir. **EDIBILITY:** Edible but of poor quality.

Suillus lakei

This attractive mushroom is readily recognized by the striking combination of reddish fibrillose scales on a yellow background, ringed yellowish stalk, and yellow pores that bruise brown. If the reddish scales wear away in rainy weather or in age, *S. lakei* can be confused with **S. caerulescens** (p. 330), which has a yellowish streaked cap with few fibrillose scales. Both of these mushrooms fruit under Douglas-fir, and both are edible but of poor quality.

Suillus pungens

CAP: 5 to 14 cm in diameter, convex to plane, whitish becoming gray or olive gray and developing other colors, in age olive, yellowish, to reddish brown, viscid, smooth, the margin inrolled when young, flesh white but aging yellow, not bluing when bruised, odor pungent or sweet. **PORES:** White becoming yellow to brown, not bluing when bruised, exuding white droplets in moist weather. **STALK:** 3 to 7 cm long, 1 to 2 cm thick, equal, surface dry, white to yellowish with red or brown dots, not bluing when bruised, partial veil absent. **SPORES:** Brown, elliptical, smooth. **HABITAT:** Single, scattered, or gregarious under pines, especially common under Monterey pine. **EDIBILITY:** Edible but of poor quality and an unpleasant taste to some.

Suillus pungens

Several features define this common slippery jack associated with pines: the viscid to slimy cap undergoes changes in color from white to gray with olive tones and finally some shade of brown; the pores of young specimens exude a milky latex; and it has a pungent odor, as the name suggests. **Suillus granulatus** (p. 331) has an orange-brown to cinnamon brown viscid cap, yellow pores, and a yellowish stalk with reddish brown glands. Like *S. pungens*, it lacks a partial veil. It is common and widespread under a variety of pines.

Suillus tomentosus

CAP: 5 to 12 cm in diameter, convex, yellowish to pale orange-tan with grayish brown or reddish fibrils or scales but often becoming smooth in age, the surface viscid when moist (but soon dry), margin inrolled at first, flesh white to yellow, bluing slowly when bruised. **PORES:** Brownish when young becoming yellow-brown in age, angular, bruising blue. **STALK:** 4 to 11 cm long, 1 to 3 cm in diameter, equal or enlarged below, yellowish with brown glandular dots, flesh yellow, usually bluing slowly when bruised, partial veil absent. **SPORES:** Olive brown, elliptical to spindle shaped, smooth. **HABITAT:** Scattered in mycorrhizal association with two-needle pines, such as lodgepole and beach pines. **EDIBILITY:** Edible but of poor quality.

Suillus tomentosus

The fibrillose yellow cap, bluing reaction of the cap and pores when bruised or cut, and absence of a partial veil are the principal features of this mushroom commonly found under pines. **Suillus fuscotomentosus** (p. 329) is similar but does not stain blue. It is common under three-needle pines, such as Monterey pines and Ponderosa pines. **Boletus subtomentosus** (p. 326) is somewhat similar, but that species has a dry, velvety brown cap and lacks glandular dots on its stalk. It occurs under both pines and oaks and has a cap that may or may not bruise blue.

Fistulina hepatica
Beefsteak Fungus

CAP: 7 to 30 cm broad and 2 to 6 cm thick, fan shaped, reddish orange or reddish brown, gelatinous, firm and fleshy, the surface at first velvety, in age becoming nearly smooth, the margin often lobed, flesh reddish brown or paler, marbled with grayish streaks, soft, exuding a red bloodlike juice when cut, taste sour. **SPORE-BEARING SURFACE:** Individual tubes tightly packed together, white or yellowish becoming reddish in age. **STALK:** Absent or very short. **SPORES:** Pink to pinkish brown, elliptical,

Fistulina hepatica

smooth. **HABITAT:** Single or occasionally in groups near the base of hardwoods. **EDIBILITY:** Edible, although some find the sour taste disagreeable.

This unique species has discrete tubes, rather than pores, on the undersurface of the cap. The texture of the fruiting body is also unique—not unlike a slab of meat. Even the reddish color is beeflike. The gelatinous texture, marbled flesh that bleeds, and growth on the base of trees are also important field marks. It is not likely to be confused with any other species. It has a lemony flavor when cooked. Be careful when collecting it, as it will stain any container.

Polyporus varius
(Other name: *Polyporus elegans*) Black Foot
CAP: 2 to 7 cm broad, plane to centrally depressed or vase shaped, tough, round to kidney shaped, buff, tan, to yellowish brown, fading in age, the surface smooth and dry with a wavy margin, flesh thin and whitish to yellowish brown, taste mild to astringent. **SPORE-BEARING SURFACE:** Minute pores (four to six per millimeter), round or angular, white aging brown, decur-

Polyporus varius

rent, not separable from the cap. **STALK:** 0.5 to 6 cm long, 4 to 8 mm wide, central, eccentric, or lateral, equal or tapering toward the base, cream colored or tan with a black lower portion, dry, tough. **SPORES:** White, long-elliptical, smooth, inamyloid. **HABITAT:** Single or in small groups on decaying branches and logs of hardwoods. **EDIBILITY:** Inedible.

The buff or tan cap, conspicuously black base of the stalk, and growth on wood make this species easy to recognize. ***Royoporus badius*** (*Polyporus badius*) is similar, but it has a dark brown or reddish brown cap that is often larger (4 to 20 cm in diameter) than the cap of *P. varius*, and a stalk that is often completely black or at least brown at the top and black at the base. It grows on decaying wood of hardwoods and conifers. ***Coltricia cinnamomea***, which grows on the ground, has a thin, cinnamon, fibrillose, zoned cap (1 to 5 cm broad) with brown flesh and a rusty brown stalk.

Polyporus tuberaster
Stone Fungus

CAP: 4 to 15 cm in diameter, convex to plane and centrally depressed to funnel shaped, yellowish to yellow-brown to brown with brown fibrils or radiating fibrillose scales, dry, the margin often inrolled, flesh whitish. **SPORE-BEARING SURFACE:** Decurrent

Polyporus tuberaster

tiny pores (one to three per millimeter), white to buff. **STALK:** 10 cm or more long, 2 to 4 cm wide, equal, central or eccentric, whitish, buff, or colored like the cap, tough, growing from a brown to black underground sclerotium (a firm rounded mass of hyphae intermixed with soil, in this case) that varies from smaller than a potato to much larger. **SPORES:** White, long-elliptical, smooth. **HABITAT:** Single or in small numbers on ground and sometimes on rotten wood in hardwood/coniferous forests; widespread but not common. **EDIBILITY:** Edible but tough.

The yellow-brown to brown cap of this stalked polypore and the large underground "tuber" makes this mushroom unique. It fruits on the ground or grows through decayed wood and fruits above ground on logs and fallen branches. The sclerotium gives rise to the fruiting body and probably functions as a nutrient reservoir and a means to survive adverse conditions for extended periods.

Jahnoporus hirtus

(Other name: *Polyporus hirtus*)

CAP: 5 to 15 cm or more broad, convex to plane, fan shaped, dry, surface brown, covered with short stiff hairs, flesh white and firm, odor mild, taste distinctly bitter. **SPORE-BEARING SURFACE:** Whitish pores (one to two per millimeter), round at first but

Jahnoporus hirtus

soon angular, adnate to decurrent. **STALK:** 2 to 10 cm long, 1 to 3 cm wide, eccentric or lateral, equal or tapering toward the base, brown, covered with short hairs like the cap, solid and firm. **SPORES:** White, spindle shaped, smooth, inamyloid. **HABITAT:** Single or scattered on the ground in hardwood/coniferous forests. **EDIBILITY:** Inedible.

This distinctive polypore is characterized by the brown velvety cap, off-center stalk, and unpleasant bitter taste. ***Bondarzewia mesenterica*** (*B. montana*) is similar, but it usually grows in large compound clusters (many overlapping caps sharing the same rooting base) up to 80 cm across. Individual caps are 5 to 25 cm broad, fan shaped, tan to ochre to brown, with a velvety surface. The pores are white, angular, decurrent, and relatively large (0.5 to 2 mm across). The spores are white or creamy, globose or subglobose, and ornamented with amyloid warts and ridges. It grows at the base of conifers in the mountains.

Albatrellus ovinus
Sheep Polypore

CAP: 4 to 20 cm in diameter, convex becoming plane or centrally depressed in age, white becoming dull yellow or tan, sometimes with pink tints, the surface dry, dull, and smooth at first but usually developing cracks in age, the margin often lobed, flesh firm, whitish to yellowish, odor and taste mild. **PORES:** Small (two to four per millimeter), circular to angular, decurrent, white aging or bruising yellowish. **STALK:** 3 to 10 cm long, 1 to 4 cm wide, central or eccentric, sometimes branched, equal or enlarged toward the base, solid, firm, whitish. **SPORES:** White, subglobose to elliptical, smooth, inamyloid. **HABITAT:** Single, gregarious, or in clusters in mycorrhizal association with conifers. **EDIBILITY:** Apparently edible if thoroughly cooked but not recommended.

The principal field marks of this terrestrial polypore are the dull white to yellowish cap that becomes cracked in age, tiny white pores, and well-developed stalk. Similar species include **Scutiger pes-caprae** (*Albatrellus pes-caprae*), with a fibrillose-scaly, reddish brown to blackish brown cap 5 to 20 cm in diameter. It has relatively wide, angular, white pores (1 to 2 mm in diameter) that may be tinged pink. Its mild taste distinguishes it from the bitter-tasting **Jahnoporus hirtus** (p. 338). **Scutiger ellisii** (*Albatrellus ellisii*) has a coarsely scaly yellow-green to yellow-brown cap and relatively wide pores that stain green when bruised; **Albatrellopsis flettii** (*Albatrellus flettii*) has a beautiful light blue cap, white stalk, and white pores.

Albatrellus ovinus

Cryptoporus volvatus

Cryptoporus volvatus

FRUITING BODY: A spherical or oval hollow ball, firm and woody in age, 2 to 6 cm in diameter, cream colored to tan; an access hole is made by insects or a tear eventually forms on the under-surface. **PORES:** Three to four per millimeter and up to 5 mm long, white to tan, attached to the upper inner surface of the fruiting body, revealed when the fruiting body is cut lengthwise. **STALK:** Absent. **SPORES:** Pinkish, cylindrical, smooth. **HABITAT:** Single or commonly scattered on recently dead standing or fallen conifers. **EDIBILITY:** Inedible.

This common wood rotter is easily recognized by its unique shape—an enclosed ball encasing a layer of tubes. Wood-boring beetles and other insects eventually gain access by chewing through the bottom of the globelike fruiting body. As the insects feed on the spores and tubes, they gather spores that are then carried to new sites, an example of mutualism—the insects are housed and fed by the fungus, and the fungus is spread from one tree to another by the mobile insects, which bore through the bark of trees, providing a new home for the fungus.

Antrodia xantha

Antrodia xantha

FRUITING BODY: Resupinate and tightly attached to the substrate, a few millimeters thick and up to a meter long, smooth or sometimes slightly raised, white aging cream colored to pale yellow, soft and spongy when fresh, brittle and chalky when dry, taste bitter. **PORES:** Round or angular, four to six per millimeter, tube length 3 to 5 mm, margin of the fruiting body without pores. **STALK:** Absent. **SPORES:** Colorless, long-elliptical, curved, smooth. **HABITAT:** On dead wood of hardwoods and conifers. **EDIBILITY:** Inedible.

This polypore is quickly identified by its white color, tiny pores, and resupinate habit. Microscopic analysis is required, however, to distinguish it from the many other species that are macroscopically similar. Examples include **Skeletocutis amorpha,** which is fully resupinate or effused-reflexed and up to 2 mm thick. The cap margin, when present, is finely hairy and grayish white. The thin fertile surface consists of pale salmon or pinkish orange, round or angular pores (three to four per millimeter). It grows on dead wood of conifers. These fungi are known as brown rot fungi, which destroy wood by degrading cellulose without extensively changing the lignin. Logs that have been decomposed by brown rot fungi fall to pieces as small, brown, shrunken cubes.

Pycnoporellus alboluteus

FRUITING BODY: A sessile, slightly raised cap or more commonly resupinate mass of pores, the fruiting body spongy, soft, and lightweight, often pitted, the surface a mixture of light orange to apricot and white colors, annual; an individual fruiting body may be up to 3 cm thick and more than 15 cm across but fruiting bodies coalesce and a fructification may be up to a meter long, odor fragrant. **PORES:** Up to several millimeters in diameter, round or more commonly angular, becoming toothlike and pointed as the walls wear away, flesh soft and pale orange. **SPORES:** Hyaline, cylindrical, smooth. **HABITAT:** On fallen logs in the high mountains soon after the snow melts. **EDIBILITY:** Unknown but insubstantial.

This wood rotter is fairly common in the high mountains on fallen conifer logs and occasionally aspen. The overall orange color, toothlike edges of the pores, and soft texture are diagnostic. *Pycnoporellus fulgens* is similar, but it has a distinct cap and grows shelflike. *Pycnoporellus alboluteus* often grows in company with *Oligoporus leucospongia* (p. 354), which is also soft and spongy but is white and shell shaped. Both species occur soon after the snow melts in spring. *Ceriporia spissa*, which is orange like *P. alboluteus*, grows tightly appressed to its wood substrate and has a soft gelatinous texture.

Pycnoporellus alboluteus

Gloeophyllum sepiarium

CAP: 2 to 15 cm broad, irregularly fan shaped, rusty brown to brown with a yellow-orange margin when actively growing, dry, velvety or hairy but aging almost smooth, concentrically zoned and wrinkled, annual, flesh thin, brownish, leathery and tough, odor and taste mild. **PORES:** Gill-like but sometimes forming elongated pores, yellow-brown to brown. **STALK:** Absent. **SPORES:** White, cylindrical, smooth. **HABITAT:** Single, in clusters, or in overlapping tiers of caps on decaying conifer wood, causing a brown rot of the sapwood and heartwood. **EDIBILITY:** Inedible.

The hairy, zoned, rusty brown cap with a yellow-orange growing margin, brown "gills" or elongated pores, and brown flesh are the important field marks for this unusual fungus. It is common on dead conifers in the mountains of the West. *Lenzites betulina* is similar, but it has a white to gray hairy cap with concentric zones and whitish to buff gill-like pores. The surface of the cap apparently retains water because older caps are often green from algal growth. Viewed from above, it has the appearance of *Trametes versicolor* and relatives (p. 356), but the gill-like pores distinguish it. It is common on hardwood logs and stumps. The cap of *Daedalea quercina* is whitish to tan or brown and concentrically zoned, the pores are whitish, irregular, and mazelike rather than gill-like, and the flesh is whitish to pale brown. It grows on hardwoods, especially oak.

Gloeophyllum sepiarium

Conks

This is a group of wood decay fungi that are not necessarily closely related. The fruiting bodies, or conks, are a common sight on dead trees and logs, jutting out at right angles from the wood. The conks are typically fan or hoof shaped and sometimes grow shelflike in tiers. The common features of this group of species are the pores, which are not separable from the cap, and the tough consistency of the fruiting body, which does not readily decay like fleshy mushrooms.

a. Pore surface pink.......................................**Fomitopsis cajanderi**

a. Pore surface not pink
    b. Cap surface with a varnished crust, at least in part; growing margin whitish
        c. Flesh punky, cap brownish, on hardwoods........................
        ....**Ganoderma lucidum**
        c. Flesh punky, cap brownish, on conifers ....
        ....**Ganoderma oregonense** (see *G. lucidum*)
        c. Flesh hard and woody, cap brown and red ....
        ....**Fomitopsis pinicola**
    b. Cap surface not varnished
        d. Flesh white or yellowish; cap yellow to orange.................
        .... **Laetiporus conifericola**

Ganoderma oregonense

  d. Flesh brown

    e. Pore surface white, instantly bruising brown ....
    ....***Ganoderma applanatum***

    e. Pore surface brownish, at least in age

      f. Fruiting on the ground; cap usually compound.......
      ....***Phaeolus schweinitzii***

      f. Fruiting on wood; cap solitary but often in rows or
      columns.. ***Phellinus pini***

Fomitopsis cajanderi

CAP: 3 to 10 cm broad, up to 2 cm thick, growing shelflike, hoof shaped or irregularly semicircular, plane to convex, pink or pinkish brown aging reddish brown and finally almost black, the surface minutely hairy becoming smooth or roughened and faintly concentrically furrowed, leathery becoming tough and hard in age (but not encrusted), margin thin and pink when actively growing but later thicker and lobed, annual or short-lived perennial, flesh pink to pinkish brown, odor and taste mild. **PORES:** Tiny (three to four per millimeter), circular, pink when young, becoming pinkish brown in age. **STALK:** Absent. **SPORES:** White, cylindrical and slightly curved, smooth. **HABITAT:** Shelf-like, scattered, or in overlapping tiers on dead conifers, often at the cut ends of logs, causing a brown rot. **EDIBILITY:** Inedible.

Fomitopsis cajanderi

The pink coloration of the fruiting body, especially the pink undersurface, is distinctive. *Fomitopsis pinicola* (below) has a red band on the top of the cap, and the pore surface is white (or yellow or brown in age but never pink). Other conks with reddish colors on the cap have a varnished surface (see **Ganoderma lucidum** [p. 349] and **G. oregonense** [p. 345]). *Oligoporus placentus* is pink and completely resupinate and rots conifer wood.

Fomitopsis pinicola
Red-belted Conk

CAP: 5 to 30 cm or more broad, up to 15 cm or more thick, growing shelflike, rounded, hoof shaped, or semicircular in outline, hard and woody, at first yellowish brown, becoming dark brown to almost blackish, concentrically furrowed in age, with a red band behind the white growing margin, which is thick and rounded, the surface developing a resinous varnished crust, perennial, flesh buff to brown, bruising pinkish when young, tough and woody. **PORES:** Tiny (three to five per millimeter), distinctly layered, white bruising yellowish and aging brownish, sometimes exuding droplets of water. **STALK:** Absent. **SPORES:** White or pale yellow, cylindrical to long-elliptical, smooth. **HABI-**

Fomitopsis pinicola

TAT: Single or scattered on dead trees, logs, and stumps or sometimes living trees, primarily on conifers, causing a brown rot. **EDIBILITY:** Inedible.

Fomitopsis pinicola is a very common conk recognized by the red, varnished band behind the white growing margin and the substantial size of the cap. Unlike **Ganoderma lucidum** (p. 349), the tissue is hard and woody, not corky and punky. **Laricifomes officinalis** (*Fomitopsis officinalis*) has a conspicuously hoof-shaped and relatively tall (cylindrical), whitish or gray cap, whitish flesh, and a very bitter taste. It is not uncommon on conifers, but it typically fruits high on tree trunks, so it is more often encountered on fallen trees.

Ganoderma applanatum

Artist's Conk

CAP: 5 to 50 cm or more broad, 2 to 10 cm thick, fan shaped or semicircular in outline, growing shelflike, the surface dull brown, unvarnished, hard and woody, furrowed (concentric growth rings visible on older specimens), perennial, flesh brown, up to 10 cm thick but typically less than 4 cm. **PORES:** Tiny (four to six per millimeter), white, immediately and permanently bruising brown, distinctly layered (a layer of tubes up to 1 cm long is added each year). **STALK:** Absent. **SPORES:** Brown, elliptical and blunt on one end, double walled with connectors. **HABITAT:** Single or, less commonly, in small groups on hardwood and coniferous stumps, logs, and living trees, especially in partially hollowed areas near the base of the tree, causing a white rot. **EDIBILITY:** Inedible.

The brown color of the zonate cap, often dusted with spores that have been carried by air currents from the pores, perennial growth, and the white pore surface that bruises brown are a distinctive set of field marks. The immediate bruising reaction permits a person to draw on the underside, hence the name Artist's Conk (inset). **Ganoderma brownii** is similar, but it has a taller cap (flesh up to 20 cm thick). The pore surface of **G. oregonense** (p. 345) and **G. lucidum** (p. 349) may bruise brown when fresh, but these annual species have varnished reddish brown caps. **Fomitopsis pinicola** (p. 347) is a perennial species with a reddish brown band near the cap margin and pores that bruise yellowish.

Ganoderma applanatum, with inset photo of white surface bruised brown.

Ganoderma lucidum
Reishi, Ling Chi

CAP: 6 to 30 cm broad, 2 to 6 cm thick, kidney shaped, circular, or semicircular in outline, concentrically furrowed as it grows in spurts, bright reddish brown aging dark brown to black, the actively growing margin white, the surface varnished but becoming dull in age, annual, flesh punky. **PORES:** Tiny (four to seven per millimeter), white, bruising brown when fresh, aging brown. **STALK:** Lateral, almost central, or sometimes absent, colored like cap, varnished, irregularly shaped. **SPORES:** Brown, elliptical, smooth. **HABITAT:** Single or in small clusters, shelflike at the base of hardwood trees and shrubs or from roots remaining in the ground when a tree has been removed, in woods and in landscaped areas. **EDIBILITY:** Inedible but used to make a bitter tea and to make various medications and potions.

The varnished reddish brown, relatively thin cap, brown spores, and varnished stalk (when it is present) define this attractive conk that causes a white rot of its host. It is commonly cultivated and used in Eastern medicine to improve general vitality. For many it is the medicinal mushroom of choice and is sometimes called the "Mushroom of Immortality." ***Ganoderma oregonense*** (p. 345) is similar but larger (especially thicker) and grows on conifers. Like *G. lucidum*, the fruiting bodies are annual.

Ganoderma lucidum

Laetiporus conifericola

(Other name: *Laetiporus sulphureus*) Chicken-of-the-Woods,
Sulfur Shelf

CAP: 5 to 60 cm broad, up to 4 cm thick, fan shaped or semicircular in outline, the surface suedelike becoming smooth, wrinkled, a mixture of bright yellow and orange, the growing margin yellow, losing the bright color in age and finally becoming pale buff, annual, flesh soft and watery when young, becoming tough and brittle in age, pale yellow, odor mild, taste mild or sour.
PORES: Tiny, two to four per millimeter, bright yellow when young, fading in age. **STALK:** Absent. **SPORES:** White, oval, smooth. **HABITAT:** Shelflike in clusters on dead and living conifer trees, stumps, and logs, often fruiting early in the mushroom season. **EDIBILITY:** The tender parts are edible, but some people experience gastrointestinal upsets.

This common species is easily recognized by the bright yellow and orange colors of the top and underside of the cap, growth on wood, and clustered habit. It often fruits from the same tree or stump year after year, causing a brown rot. The texture of cooked pieces is similar to chicken meat, thus the common name. Two species occur in the West: *L. conifericola* on conifers and **L. gilbertsonii** on hardwoods and eucalyptus. Some

*Laetiporus
conifericola*

people react adversely after eating specimens from eucalyptus. ***Laetiporus sulphureus*** is similar but grows east of the Rocky Mountains.

Phaeolus schweinitzii
Dyer's Polypore
CAP: Up to 25 cm or more in diameter, circular or fan shaped, several caps typically growing from a single base, spongy and knobby with a hairy, creamy, greenish yellow, or orange margin when young, becoming dark brown, corky, and woody in age, flesh yellow-brown aging dark brown. **PORES:** Decurrent, greenish yellow but soon brown. **STALK:** Short (up to a few cm), rooting, central or eccentric, colored like cap. **SPORES:** White, elliptical, smooth, inamyloid. **HABITAT:** Single or in groups on the ground at or near the base of dead and living conifers, causing a brown rot of its host. **EDIBILITY:** Inedible.

Phaeolus schweinitzii

Although an annual, the fruiting bodies persist from one year to the next, and old, brown, compound fruiting bodies at the base of conifers are frequently encountered. Other conks with hairy caps include **Onnia tomentosa** (*Inonotus tomentosus*), which is smaller and duller in color than *P. schweinitzii* and typically grows as individual fruiting bodies on the ground or shelflike on conifers; **Inonotus dryophilus**, which grows shelflike on oaks; and **Heterobasidion irregulare** (*Heterobasidion annosum*),which is an annual or perennial with whitish pores and white flesh that grows at or near the base of living pine and juniper trunks, causing a white rot of commercial importance—it is highly variable, sometimes growing flat on the surface of wood and other times having a knobby and irregularly shaped brownish cap with a lighter growing margin. The similar **H. occidentale** has a preference for fir and hemlock.

Phellinus pini

(Other name: *Porodaedalea pini*)

CAP: 2 to 20 cm broad, 2 to 15 cm thick, growing shelflike, fan or hoof shaped, woody, the surface rough, crusty, and concentrically furrowed, reddish brown with a brighter growing margin, aging blackish, perennial, flesh reddish or cinnamon brown. **PORES:** Variable in size and shape, some longer than wide and curved, others angular, cinnamon brown. **STALK:** Absent. **SPORES:** Brown, subglobose, smooth, inamyloid. **HABITAT:** Single, in groups, or in columns on trunks of living or dead conifers, causing a white rot of heartwood. **EDIBILITY:** Inedible.

This important pathogen of conifers causes serious economic losses for the timber industry. It is characterized by the woody, rough, reddish brown cap, irregular pores, and brownish flesh. *Phellinus gilvus* is similar but has tiny circular pores (five to seven per millimeter), a thinner cap (1 to 3 cm), a yellow growing margin, and preference for oaks. The cap of *P. igniarius* is up to 40 cm broad and 20 thick, hoof shaped, hard and woody, and blackish in age with a cracked surface and has tiny circular pores and a sour or bitter taste. It grows on hardwoods. *Phellinus tremulae* is also blackish in age and grows on aspen. Many other *Phellinus* species grow on wood with a resupinate growth habit, such as *P. ferruginosus*, which grows on logs of hardwoods and conifers. Like all *Phellinus* species, *P. ferruginosus* is characterized by brownish flesh that darkens in potassium hydroxide, a dark pore surface, and the presence of microscopic setae (tiny stiff hairs) in the pores.

Phellinus pini

Oligoporus leucospongia

(Other name: *Tyromyces leucospongia*)

CAP: Fan shaped and shelflike or almost resupinate, often longer than wide (4 to 12 cm long and 1 to 5 cm wide), convex, white or white tinged tan, soft and cottony, the cap surface velvety becoming smooth and firmer than the flesh, the margin of the surface thin and hanging over the pore surface, annual, flesh white, 1 to 2 cm thick, odor mild, taste acrid. **PORES:** Two to three per millimeter, up to 6 mm long, round or angular, becoming ragged edged and toothlike in age, whitish aging buff. **STALK:** Absent. **SPORES:** White, sausage shaped, smooth, inamyloid. **HABITAT:** Single or scattered on dead conifers in the mountains soon after the snow melts but found year-round since the fruiting bodies are slow to decay. **EDIBILITY:** Inedible.

The soft cottony or foamlike texture, rounded growing margin that hangs over the pore surface, overall white color, and growth on wood characterize this species. **Oligoporus fragilis** has a white cap and pore surface that bruises yellowish and then reddish brown. It is soft and watery when fresh and brittle when dry. **Leptoporus mollis** is similar, but it bruises reddish brown directly. **Oligoporus caesius** (*Tyromyces caesius*) is also soft and spongy. It has a whitish cap with faint blue tones and a fragrant

Oligoporus leucospongia

odor. **Tyromyces chioneus** is whitish overall without any staining reactions. It has a soft watery texture and a fragrant odor.
Oligoporus amarus (*Tyromyces amarus*), which is restricted to
incense cedar, has a whitish cap and a yellow pore surface. The
fruiting body becomes hard and rigid with age.

Trichaptum abietinum

CAP: 1 to 4 cm broad, up to 0.5 cm thick, fanlike or shelflike or
sometimes flat, dry, hairy, white or gray, faintly concentrically
zoned, margin purplish and wavy, annual or short-lived perennial, flesh thin, leathery, brownish. **PORES:** Tiny, two to four per
millimeter, shallow, purplish becoming brownish with a pale
purplish tinge in age, the walls breaking up unevenly and finally
appearing toothlike and irregular in size and shape. **STALK:** Absent. **SPORES:** White, cylindrical, smooth. **HABITAT:** Shelflike in
clusters, rows, or overlapping tiers or flat on rotting coniferous
wood. **EDIBILITY:** Inedible.

The purplish tinted cap margin and pore surface, ragged-
edged pores, and growth on conifers are the important field
marks of this abundant wood rotter. **Trichaptum biforme** is similar but is somewhat larger and grows on hardwood stumps and
logs. **Trametes versicolor,** the Turkey Tail (p. 356), has a distinctly zoned, multicolored cap and lacks the purplish tints on

Trichaptum abietinum

the undersurface. ***Bjerkandera adusta*** has grayish to brownish, hairy, thin caps with a leathery texture and minute, smoky gray, angular pores that bruise or age darker. It usually occurs in clusters and overlapping tiers on decaying deciduous wood and occasionally on coniferous wood but sometimes grows flat on its substrate. It causes a white rot.

Trametes versicolor
Turkey Tail

CAP: 2 to 10 cm broad, shelflike and fan shaped, annual, dry, velvety, concentrically zoned with alternating textures and brown, reddish brown, yellow, and gray colors, margin wavy, flesh thin, tough, and leathery. **PORES:** Tiny (three to five per millimeter), circular, whitish. **STALK:** Absent, the fruiting body broadly or narrowly attached to the substrate. **SPORES:** White, cylindrical, smooth. **HABITAT:** Shelflike in rosettes, rows, or clusters on logs and stumps of dead hardwoods. **EDIBILITY:** Inedible but used to make a medicinal tea.

This abundant species is readily identified by the turkey-tail-like coloration of the cap due to alternating bands of color. The thin tough caps do not readily decay, so fruiting bodies are

Trametes versicolor

found year-round. **Trametes hirsuta** is similar, but the white to gray cap is densely hairy, only faintly concentrically zoned, and usually has a brown margin. It is very similar to **Lenzites betulina** (p. 344), but that species has large platelike gills. **Trametes suaveolens** has whitish or pale brown pores that bruise brown, a thicker (up to 1 cm), hairy, azonate cap that becomes smooth in age, and an odor of anise. It also grows on hardwoods. Also see **Bjerkandera adusta** (p. 356). These fungi may resemble **Stereum hirsutum** (p. 311), but the spore-bearing surface of that species is smooth, and the cap is hairy and orange-brown.

Nidula candida

FRUITING BODY: A small cup-shaped "nest" 5 to 15 mm high, cylindrical to cushion shaped with a flaring mouth at maturity, the outer surface tawny or dull yellow-brown and covered with hairs, the nest at first closed by a hairy lid, which breaks away, revealing the smooth, cream-colored to buff inner surface and several buff to brown, flattened peridioles ("eggs") about 2 mm in diameter embedded in a sticky gel. **SPORES:** Hyaline (colorless), elliptical, smooth, borne inside the peridioles. **HABITAT:** Scattered to gregarious on sticks and woody debris. **EDIBILITY:** Inedible.

Nidula candida

The eggs of bird's nest fungi are forcefully splashed out of the nest by rain drops. In this species, the eggs are embedded in a mucilaginous gel; in some others, such as **Crucibulum laeve,** the eggs are attached to cords that wrap around twigs and other plant material when the eggs (and, therefore, the spores) are splashed out of the nest. In *C. laeve*, the eggs are whitish and the velvety exterior of the nest is tawny or dull yellow-brown. In **Cyathus stercoreus,** flattened and blackish eggs are attached to the nest with a short cord. It has a cone-shaped nest with a circular mouth, a brown hairy exterior that becomes smooth in age, and gray interior. In **Cyathus olla,** the mouth of the gray, minutely hairy nest is often wavy and the peridioles are gray. The interior wall of the cup of **Cyathus striatus** is conspicuously grooved.

Phallus impudicus
Stinkhorn

FRUITING BODY: In maturity, an erect cylindrical stalk with a foul-smelling, spore-bearing cap arising from a buried "egg." **CAP:** Up to 4 cm wide, white with distinct ridges, coated with a greenish mucilaginous layer of spores, which has a very disagreeable putrid odor, often with an opening at the tip. **STALK:** Up to 15 cm or more long and 2 to 4 cm wide, white, hollow, tubular, and fragile. **VOLVA:** White, saclike, with thick mycelial cords often present at the base. **HABITAT:** Single or in groups in lawns, gardens, other landscaped areas, and wood chips. **EDIBILITY:** Edible in the egg stage; otherwise, inedible due to the putrid odor.

This unique fungus starts as an oft-unnoticed "egg," which soon ruptures as the stalk elongates, elevating the slimy spore mass above the substrate. At this stage, flies are attracted to the odor and carry spores away as they visit the fruiting body. **Phallus hadriani** is identical except the volva is tinted purplish. Other stinkhorns share the same attraction for insects but are often more ornate. **Lysurus cruciatus** has thick, erect, red, spore-bearing arms on top of a white cylindrical stalk. The arms separate at maturity. The spore mass, which has a putrid odor, is olive brown. It grows in lawns, gardens, and other landscaped areas. **Lysurus mokusin** is similar, but the red arms usually remain touching, and the stalk is star shaped in cross section. **Clathrus ruber** has a red, spherical, latticelike configuration and lacks a stalk.

Phallus impudicus

Tulostoma campestre
Stalked Puffball

SPORE CASE: Spherical, 1 to 2 cm in diameter, consisting of two layers; in maturity, the outer layer breaks away, leaving a sand-encrusted cup around the inner layer, which remains as a buff to yellow-brown spherical spore case with a poorly defined tube around the apical pore. **STALK:** Up to 6 cm long, 2 to 4 mm wide, dry, buff to brown. **SPORES:** Brown, subglobose to elliptical, ornamented with small blunt warts. **HABITAT:** Single or small groups in sandy soil in arid regions of the West. **EDIBILITY:** Inedible.

This is one of many stalked puffballs. As a group, they prefer dry areas and nutrient-poor soils, including deserts of the Southwest. ***Battarrea phalloides*** is a much taller and larger-stalked puffball. The long, thin, scaly or ribbed stalk is up to 40 cm long and has a large volva (which may go unnoticed in the ground). The spore case, which is up to 6 cm in diameter, splits in maturity and falls off, exposing the spore mass. Once the

Tulostoma campestre

spores are dispersed by the wind, the stalk and remaining disk-like bottom of the spore case persist. It is widespread, including desert and arid regions.

Agaricus deserticola

(Other names: *Agaricus texensis, Longula texensis*) Gastroid Agaricus

CAP: 4 to 10 cm in diameter and 4 to 10 cm in height, oval or spherical, often with a flattened top, dry, white, smooth but soon developing scattered white to brown scales, the cap completely enclosing cavities containing the gills, fragile and brittle in age and splitting open, releasing the spores, flesh bruising yellow. **GILLS:** Contorted and convoluted dense plates, free. **SPORES:** Chocolate brown, globose to elliptical, smooth, not forcibly discharged. **STALK:** 4 to 10 cm long, 2 to 4 cm wide, equal or wide at the base, white, reaching the top of the cap (running through the spore cavity), partial veil leaving a ring on the stalk (if it separates from the stalk). **HABITAT:** Single or scattered on ground in disturbed areas, cultivated fields, vacant lots, and so forth, common in arid areas but widespread. **EDIBILITY:** Unknown.

The chocolate brown spores, white scaly cap that remains closed, and brittleness of the fruiting body in age are important

Agaricus deserticola

field marks. The cap and stalk stain yellow, at least in young specimens. Except for the failure of the cap to open, it clearly reminds one of a species of *Agaricus*. **Agaricus inapertus** (*Endoptychum depressum*) is similar, but the cap is typically smooth, the stalk is short, and it grows under conifers in the mountains. Like *A. deserticola*, it also bruises yellow.

Puffballs and Earthstars

The puffballs and earthstars produce spores in an enclosed chamber, whereas most fungi covered in this field guide produce spores on the surfaces of gills, spines, branches, or pores. Because the spores are formed inside a closed structure, they have lost the ability to be ejected from the basidia. Thus, a spore print is not obtainable. In the earthstars, the outer wall or peridium of the fruiting body opens in a stellate (starlike) fashion, exposing the inner wall, which has an apical pore.

a. Outer layer splitting into starlike rays that unfold
    b. Rays hygroscopic (unfolding when moist)............................
    .... **Astraeus hygrometricus**

    b. Rays not hygroscopic (permanently unfolded)
        c. Rays upright and elevating the spore case ....
        ....**Geastrum fornicatum**
        c. Rays flat or folded under the spore case.............................
        ....**Geastrum saccatum**

a. Outer layer generally not splitting into rays or, if so, the rays tough
    d. Spores mass containing small chambers ....
    ....**Pisolithus arhizus**
    d. Small chambers absent; spores produced in a single large chamber
        e. Spore case tough and thick walled; spore mass firm and purple-black when young..........**Scleroderma polyrhizum**
        e. Spore case thin walled; spore mass white when young
           f. Sterile base present (i.e., spores absent in base)
               g. Fruiting body more-or-less round, rupturing irregularly in old age
                  h. Fruiting body covered with large conspicuous warts ....**Calvatia sculpta**
                  h. Fruiting body smooth or covered with flattened scales or warts
                     i. Mature spore mass purple.................................
                     ....**Calvatia cyathiformis**
                     i. Mature spore mass olive brown.......................
                     ....**Lycoperdon utriforme**
               g. Fruiting body pear shaped, tearing to form an apical pore
                  j. Growing on the ground; exterior with cone-shaped spines (leaving scars when they fall off in age)................................ **Lycoperdon perlatum**
                  j. Growing on wood or lignin-rich soil; exterior smooth or covered with very short spines...........
                  .... **Lycoperdon pyriforme**
             f. Sterile base absent or rudimentary
               k. Fruiting in the mountains soon after the snow melts; fruiting body white with brown-tipped scales...**Calvatia subcretacea**
               k. Widespread; fruiting body not scaly
                   l. Smooth outer layer peeling away, inner layer pa-

per thin, opening through a large pore; in grass
....***Bovista pila***

l. Bottom part of the outer layer peeling away,
the upper part covered with sand and soil, thus
acornlike ....***Disciseda candida***

Astraeus hygrometricus

FRUITING BODY: 4 to 9 cm in diameter, initially subterranean and nearly spherical, at maturity the outer layers split into 5 to 15 triangular, leathery rays that fold backward when moist to expose the spherical, sessile, brown, soft spore sac (1 to 2 cm broad), the outer surfaces of the rays smooth and buff to dark brown, inner surfaces of the rays gelatinous, gray to brown, often crisscrossed with cracks when dry, spores released through a slit or pore on the top of the spore sac. **STERILE BASE:** Absent. **SPORES:** Subglobose, brown, warty. **HABITAT:** Single to gregarious, found in a variety of habitats, especially sandy or disturbed soil. **EDIBILITY:** Inedible.

When open, the fruiting body has a starlike appearance with the spore sac in the center. The rays of this earth star are hygro-

Astraeus hygrometricus

scopic—the gelatinous layer in the rays quickly absorbs water and in a matter of minutes the rays open, exposing the spore sac to rain drops. The force of raindrops disperses the spores. The fruiting body may remain functional for years, with the rays curled over the spore case, protecting the spores from the hot sun and drought and then opening to release spores when conditions are suitable. **Astraeus pteridis** is a large (up to 15 cm broad) hygroscopic earthstar with thick rays that crack abundantly in age.

Bovista pila

FRUITING BODY: Spherical to egg shaped, 3 to 9 cm in diameter, the outer layer of the peridium (skin) smooth and white but aging brown, gradually peeling away in age and exposing the papery thin inner layer, which is gray and shiny metallic or brown from patches of the outer layer, tearing away on the top to form cracks or irregular pores; eventually, the fruiting body dries and blows away, scattering spores as it rolls along the ground; odor and taste mild. **STERILE BASE:** Absent; a mycelial cord at the base of the stalk attaches the fruiting body to its substrate. **SPORES:** White becoming dark brown, globose, smooth, with a short, hyaline pedicel (stalklike appendage). **HABITAT:** Single or more often scattered in grassy areas in natural areas such as open woods and landscaped areas such as lawns and parks. **EDIBILITY:** Edible when young, firm, and white inside.

Bovista pila

The small size, metallic luster of the inner layer, and the white mycelial cord that attaches the fruiting body to the substrate are the key features. *Bovista plumbea* is similar, but the spores are oval with a longer pedicel, the inner layer of the peridium is grayish, the spores are released through an apical pore, and the fruiting body is attached to the ground by mycelial fibers, not a simple cord.

Calvatia cyathiformis
Lilac Puffball

FRUITING BODY: 5 to 20 cm in diameter, subglobose to pear shaped and often flattened at the top, furrowed below as the fruiting body is constricted to a blunt rootlike base, the outer layer white at first, becoming pale brown to purple in age, smooth to felty when young, eventually breaking up into flattened warts, scales, or starlike patches of flat purplish brown scales; the outer layer eventually falling away to expose the smooth, purple to purple-brown, paperlike inner peridium which eventually fractures, releasing the spores. **STERILE BASE:** Persistent, white at first becoming purplish, large in some specimens, rudimentary in others. **SPORES:** Borne in a cottony or powdery mass that is white to ochre when young, becoming purple at maturity, spores globose, roughened. **HABITAT:** Scat-

Calvatia cyathiformis

tered or gregarious in grassy fields, lawns, pastures, gardens, road edges, and other disturbed areas. **EDIBILITY:** Edible when the spore mass is young, firm, and white.

This puffball is easily recognized by the purplish spore mass at maturity. It is often encountered when just the purplish sterile base and lower parts of the fruiting body remain. *Lycoperdon utriforme* (p. 373) has a flattened spore case and large sterile base and has olive brown spores at maturity.

Calvatia sculpta

FRUITING BODY: Softball size or larger (up to 15 cm in diameter) with distinctive, pyramidal, terraced warts that are often curved at the tip; the wall of the spore case eventually cracks and breaks apart between the warts, releasing the spores (i.e., no apical pore). **STERILE BASE:** Sometimes well developed, attached to the substrate with white rhizomorphs. **SPORES:** White becoming olive brown, globose to subglobose, minutely spiny. **HABITAT:** Single or in groups in mountains throughout the West. **EDIBILITY:** Edible when the spore mass is young, firm, and white.

The large, terraced, pointed pyramidal warts give this puffball the appearance of a Christmas tree ornament. If collected for the table, the resemblance to *Amanita magniverrucata* should be taken into account. The latter is possibly toxic and can be identified, even when young, by cutting the fruiting body lengthwise. Young puffballs will be uniformly white, whereas outlines of young fruiting bodies of *Amanita* species will be visible in the egg stage. Other mountain puffballs include *Calbovista subsculpta*, which is about the same size as *C. sculpta* but has flatter nonterraced warts, and *Calvatia booniana*, which has a round or elongated white spore case up to 60 cm broad with flattened darker warts or scales. It grows singly or in groups in mountains throughout the West.

Calvatia subcretacea

(Other names: *Gastropila subcretacea, Handkea subcretacea*)

FRUITING BODY: Round or flattened, 2 to 5 cm in diameter, thick walled, white with scales that are tipped with short dark brown fibrils, the scales breaking into plates at maturity, eventually dis-

Calvatia sculpta

Calvatia subcretacea

integrating to release the spores, odor and taste mild. **STERILE BASE:** Absent, the fruiting body attached to the ground by a few white mycelial cords. **SPORES:** White but soon dark olive brown and then brown, globose, roughened. **HABITAT:** Single or scattered in high mountain forests, often growing soon after the snow melts. **EDIBILITY:** Edible when young, firm, and white inside.

The bright white surface decorated with dark-tipped scales gives this puffball visual appeal. It often fruits soon after snowmelt with *C. fumosa* (*Gastropila fumosa*), which is also moderately sized and thick walled, but *C. fumosa* is gray to grayish brown in color and smooth (but soon superficially cracked). A sterile base is absent in *C. fumosa*; a single mycelial cord attaches the fruiting body to the ground. The thick wall of *C. fumosa* is reminiscent of *Scleroderma*, but the grayish color of the cap of *C. fumosa* is characteristic; in addition, when young, the spore case of *Scleroderma* is firm and dark purple, whereas the spore mass of *Calvatia* is soft and white. Microscopic mycelial threads (capillitia) run through the spore mass of *Calvatia* species but not *Scleroderma*.

Disciseda candida
Acorn Puffball

FRUITING BODY: Spherical to depressed-spherical, 1 to 4 cm in diameter, 1 to 3 cm tall, the outer peridium (skin) sloughing off from the basal portion of the fruiting body revealing a smooth to granular, gray to brown inner peridium; the outer peridium binding sand and debris and remaining encrusted, resulting in a sand cup on the top of the puffball; a pore develops when the inner peridium breaks away from the rootlike structure that anchors the fruiting body to the soil. **STERILE BASE:** Absent. **SPORES:** White when young, becoming powdery and brown at maturity, globose, roughened. **HABITAT:** Scattered to gregarious in arid grasslands, chaparral, dry woods, and disturbed areas. **EDIBILITY:** Unknown but too small to be of value.

This common little puffball is often found as it tumbles in the wind like a lightweight acorn. The pore is formed at the base of the fruiting body, but it appears apical because the fruiting body flips over with the sand cup as its ballast. Depending on the characteristics of the soil or leaf litter where the fruiting body

Disciseda candida

grew, the sand cup may incorporate litter, sand, clay or sometimes nothing at all, resulting in a variety of appearances.

Geastrum fornicatum
Arched Earthstar

FRUITING BODY: Subterranean and nearly spherical when young, the outer layers soon splitting into three to six triangular rays, which become erect, elevating the spore case; the brown, scaly rays attached to a basal mycelial cup encrusted with soil debris, which also may split into rays; spore sac gray, brown, to nearly black, slightly hairy, elliptical to pear shaped, 1 to 4 cm in diameter, seated on a short stalk 2 to 5 mm long and 3 to 10 mm in diameter; the entire fruiting body is 2 to 10 cm high (excluding the cup) and 3 to 6 cm in diameter when the rays are fully expanded; the spores are released through an apical pore that is positioned on a conical protuberance. **STERILE BASE:** Absent. **SPORES:** Globose, brown, warty. **HABITAT:** Scattered to gregarious in leaf litter under trees, especially oaks and junipers. **EDIBILITY:** Inedible.

This distinctive species, which looks like a lunar landing craft or a human figure, is readily identified by the mature spore case perched on the tips of the erect rays. A very similar smaller

Geastrum fornicatum

Geastrum saccatum

species, **G. quadrifidum,** has an apical pore surrounded by a well-defined silky area. Another common species, **G. minimum,** has 5 to 12 rays and lacks a mycelial cup.

Geastrum saccatum
Sessile Earthstar
FRUITING BODY: Initially spherical or flattened, the outer layers soon splitting into four to eight triangular to lance shaped, smooth, buff to yellow-brown rays which eventually curve back under the spore case; the spore case 0.5 to 2 cm in diameter, smooth, sessile, buff, with an apical raised pore outlined by a circular pale zone; when the rays are fully expanded, the fruiting body is up to 5 cm broad. **STERILE BASE:** Absent. **SPORES:** Globose, brown, warted. **HABITAT:** Single or scattered on ground in hardwood/coniferous forests, landscaped areas, and open spaces. **EDIBILITY:** Inedible.

The light color of this earth star, the tendency of the rays to fold under, the pale buff color, and the faintly visible ring around the raised apical pore help identify this fungus. **Geastrum fimbriatum** is similar but darker in color and has a torn apical pore that is not encircled by a ring. **Geastrum triplex** is also similar, but it has a cup under the base of the spore case. Other nonhygroscopic earthstars include **G. fornicatum** (p. 369), which has a gray, dark brown, or blackish spore case perched on tips of its rays.

Lycoperdon perlatum
Common Puffball
FRUITING BODY: Pear shaped, up to 10 cm high and 3 to 7 cm wide, outer layer covered with conelike, whitish to brown spines 1 to 2 mm tall, which leave conspicuous circular scars when they fall off, inner layer smooth, white or pale brown, tearing to form an apical pore. **STERILE BASE:** Well developed, broad and stalklike. **SPORES:** White maturing olive brown, globose, minutely warted. **HABITAT:** Scattered, gregarious, or clustered on ground in woods or in landscaped areas. **EDIBILITY:** Edible when young, firm, and white inside.

The pear-shaped fruiting body, spiny surface (which may be smooth in old specimens), scars from spines as they fall off, and

Lycoperdon perlatum

growth on the ground are the principal field marks. ***Lycoperdon umbrinum*** is pear shaped and relatively small (less than 5 cm tall). It has short spines less than 1 mm tall on a dark outer peridium (skin). In age the spines fall off without leaving scars, resulting in a blackish and smooth puffball. It grows on the ground in woods. ***Lycoperdon marginatum*** is identified by the way the spiny outer layer of the peridium peels off the fruiting body in sheets, revealing the olive brown inner layer. It also grows on the ground in woods.

Lycoperdon pyriforme
(Other name: *Morganella pyriformis*)
FRUITING BODY: Pear shaped or rounded, up to 5 cm high and 2 to 4 cm wide, the outer layer (peridium) white or buff, aging brown, smooth or with a few tiny spines, the inner layer smooth, tearing to form an apical pore from which spores are released by wind and rain. **STERILE BASE:** Narrow and stalklike, white, with white rhizomorphs (ropelike strands of mycelium) attached to the base. **SPORES:** White maturing olive brown, globose, smooth. **HABITAT:** Scattered or in dense clusters on rotting wood or lignin-rich soil. **EDIBILITY:** Edible when young, white, and firm inside.

Lycoperdon pyriforme

This common puffball is characterized by its growth on decaying wood (which is often well decayed) or soil rich in woody debris. Although the fruiting bodies are covered with very tiny white spines when young, the spines soon disappear. Additional important features are the pear shape of the fruiting body (a variable characteristic), the narrow sterile base, and the white rhizomorphs. It is easily confused with **L. perlatum** (p. 371), but the outer layer of that fungus has conical spines that leave distinctive scars when they fall off. It grows on the ground.

Lycoperdon utriforme

(Other names: *Calvatia utriformis, Handkea utriformis*)

FRUITING BODY: 10 to 25 cm tall and 5 to 25 cm broad, pear shaped with a flattened top, white aging yellow-brown and finally brown, the upper part covered with fibrillose pointed scales and later, larger plates, which eventually break apart, releasing the spores. **STERILE BASE:** Well developed and large, remaining after the spore case has disintegrated. **SPORES:** White when young becoming olive brown, globose to subglobose, smooth. **HABITAT:** Single or in groups in grassy areas or open woods. **EDIBILITY:** Edible when young, firm, and white inside but of poor quality.

Lycoperdon utriforme

This puffball can be recognized by its large size, large sterile base, flattened top, and grassland habitat. Other puffballs that grow in grassy areas include **Calvatia pachyderma** (*Gastropila fragilis*), which lacks a sterile base and has a very thick wall; **C. cyathiformis** (p. 365), which has a large sterile base and purplish mature spores; **C. lycoperdoides** (*Handkea lycoperdoides*), a small whitish puffball (up to 5 cm broad) with an olive brown, cottony mature spore mass that stays intact for a while after the spore case has begun to disintegrate, and lacks a sterile base; and **Lycoperdon pratense** (*Vascellum pratense*), a small white puffball up to 5 cm broad with a sterile cuplike base, and a membrane that separates the spore mass from the sterile base.

Pisolithus arhizus

(Other name: *Pisolithus tinctorius*) Dead Man's Foot

FRUITING BODY: Rounded or pear-shaped spore case with rounded lobes, 5 to 30 cm tall and 4 to 20 cm in diameter, the thin outer layer yellow-brown to brown becoming dull brown, the interior consisting of small yellow-brown spore sacs about 3 to 5 mm in diameter that disintegrate upon maturity, releasing dusty masses of spores. **STALK:** Absent or a rooting sterile base. **SPORES:** Brown, subglobose to globose, ornamented with spines

Pisolithus arhizus

up to 2 μm long. **HABITAT:** Single or in small groups in mycorrhizal association with a variety of trees and shrubs, such as pines, oaks, and manzanitas, often fruiting with only the top part of the spore case exposed above the soil. **EDIBILITY:** Not recommended.

When mature, this fungus resembles a puffball, releasing masses of dry spores. Young specimens are unlike puffballs, however, because the spores are born in tiny chambers, easily viewed when the spore case is cut open. As the walls of the peridioles break away, the spores are released, leaving behind the sterile base. *Pisolithus arhizus* is frequently used by the forest industry to inoculate pine seedlings and establish a mycorrhizal relationship when the seedlings are transplanted into the forest. ***Dictyocephalos attenuatus*** is similar but has a smaller spore case and a long stalk with a volva. It grows in deserts and other arid areas.

Scleroderma polyrhizum
(Other name: *Scleroderma geaster*) Dead Man's Hand
SPORE CASE: Spherical, often somewhat flattened on the top, up to 15 cm broad, the peridium (skin or rind) rough, firm, and tough, 3 to 10 mm thick, white aging brown and eventually

Scleroderma polyrhizum

splitting into several rays, releasing the spores. **STALK:** Sterile base absent, attached to the ground by tough mycelial fibers. **SPORES:** White when young, then purple-black while still firm, and eventually brown and powdery when mature, globose, spiny. **HABITAT:** Single or scattered, partially buried, in poor soil, disturbed ground, and along roads. **EDIBILITY:** Avoid; some species of *Scleroderma* are poisonous.

The smooth, hard, thick peridium and purple-black coloration of the firm spore mass while immature help to separate this species from the "true" puffballs, which have a thin peridium and a white and soft spore mass while immature. The peridium of *S. polyrhizum* splits into rays in age. Relatives include **S. cepa** (p. 425), which is 2 to 8 cm broad and buff colored. The peridium, which may stain reddish when bruised, splits and cracks irregularly, exposing the spores. **Scleroderma citrinum** (*Scleroderma vulgare*) has a scaly, yellow-brown peridium that stains pinkish when cut. It grows in the woods and landscaped areas. All of these should be considered poisonous.

Rhizopogon ochraceorubens
False Truffle

FRUITING BODY: 2 to 6 cm in diameter, spherical, oval, or lumpy and irregular, yellow to ochre, overlaid with brownish rhizomorphs that stain red to reddish brown when bruised, odor fruity, taste mild. **STALK:** Absent. **SPORES:** Light gray becoming olive brown, spongy when mature, oblong, smooth, inamyloid, borne in minute chambers. **HABITAT:** Single, scattered, or gregarious, hypogeous (fruiting below ground), partially buried, or sometimes at the surface, in mycorrhizal association with pines. **EDIBILITY:** Unknown.

The yellow outer wall and adhering red-staining rhizomorphs are distinctive. This species is often only partially buried, so it is not uncommonly encountered. Members of this genus are associated with pines, like their cousin genus, *Suillus*. *Rhizopogon occidentalis* has a white to yellow outer surface and stains orange to reddish when cut or bruised. *Rhizopogon ellenae* is white when young but has lilac tints in age and stains dark purplish or black after handling. The odors of these False Truffles (as well as true truffles, which are ascomycetes, e.g., *Tuber* species, p. 384) are attractive to squirrels and other rodents, who find the fruiting bodies underground after the spores are mature. Once eaten, the spores survive the trip through the animal and are distributed in scat.

Rhizopogon ochraceorubens

Hydnangium carneum

FRUITING BODY: Subterranean, globose to elliptical or lobed and irregular in shape, 1 to 5 cm in diameter, white when young becoming pink, rose, or cinnamon, the surface velvety and smooth, pitted, or wrinkled. **STALK:** Very short or present as a white or pink, pluglike sterile base. **SPORES:** Colorless to pale yellow, globose to slightly elliptical, ornamented with spines up to 2 µm tall, borne on basidia in pink or reddish, mazelike, convoluted chambers. **HABITAT:** Single to gregarious, subterranean, mycorrhizal with eucalyptus and possibly other introduced trees and shrubs. **EDIBILITY:** Unknown.

This relative of the gilled mushroom *Laccaria* was probably introduced to America when eucalyptus was imported from Australia. Other similar false truffles include species of **Arcangeliella,** which exudes a latex when cut and has spores with amyloid ornaments, just like species of *Lactarius; Gymnomyces,* which has abundant spherical cells in the chamber walls and spores with amyloid ornaments; and **Martellia,** which has ornamented spores but typically lacks spherical cells. The latter two genera are related to species of *Russula* and are often put in that genus. See **Tuber oregonense** (p. 384), for an example of the true truffles, all of which are ascomycetes.

Hydnangium carneum

ASCOMYCETES

Annulohypoxylon thouarsianum
(Other name: *Hypoxylon thouarsianum*) Cramp Balls
FRUITING BODY: A stroma 2 to 5 cm in diameter or larger, up to 3 cm high, rounded but often flattened, tough and hard, surface black, dull or shiny, conspicuously warted, each wart containing embedded perithecia (tiny flasks where the spores are borne), surrounding each perithecial opening is a flat to sunken disklike area; the interior of the fruiting body is silky-lustrous with faint gray-brown concentric zones of radially arranged woody fibers, texture charcoal-like and brittle. **SPORES:** Brown, cylindrical, spindle shaped to curved and cucumber shaped. **HABITAT:** Scattered to clustered on the bark of fallen or standing hardwoods, especially oak. **EDIBILITY:** Inedible; much too tough and woody.

The roundish charcoal-like lumps on the bark of hardwoods are distinctive. The spore-bearing perithecia are borne in the periphery of the crustlike stroma. **Daldinia grandis** is very similar, but it lacks the flattened or sunken disk surrounding each

Annulohypoxylon thouarsianum

perithecial opening. **_Daldinia childiae_** also lacks the disks surrounding the perithecia. Internally, it has distinctive concentric zones of light and dark brown, with evidence of old perithecial cavities in some of the zones.

Xylaria hypoxylon
Carbon Antlers

FRUITING BODY: 2 to 8 cm tall and 3 to 5 mm wide, erect, tough, clublike, forked, antlerlike, or occasionally unbranched, flattened to cylindrical, initially white or grayish from spores but slowly becoming black from the base toward the tip, at least some of the tips usually remaining pale in color; perithecia (minute spore-bearing flasks) embedded in the black portion of the fruiting body, which is roughened from the protruding exit pores of the perithecia; texture leathery at first, becoming crustlike and brittle in age and when dry. **STALK:** Short, tapering upward, dark brown to black, smooth to wrinkled and hairy. **SPORES:** Dark brown to black, elliptical to bean shaped, smooth; asexual spores, which cover the fruiting body when it is young

Xylaria hypoxylon

---

and white, are colorless and spindle shaped. **HABITAT:** Scattered or in clusters on rotting wood. **EDIBILITY:** Inedible.

The blackish antlerlike fruiting body with pale branch tips and growth on wood are the principal field marks. Many similar species of *Xylaria* occur, but the taxonomy is poorly defined. One species, **Xylaria arbuscula**, has a yellow immature fruiting body. The common name used for some species of *Xylaria* is Dead Man's Fingers, referring to the similarity between *Xylaria* and fingers protruding from the ground.

Cordyceps militaris

FRUITING BODY: Erect, cylindrical to club shaped and often with a vertical furrow, 1 to 8 cm long and 2 to 10 mm in diameter, the apical portion, which is 0.5 to 5 cm long and is slightly larger in diameter than the stalk, contains embedded reddish perithecia (tiny flasks where the spores are borne) that protrude slightly to form warts, texture tough-fleshy when fresh and brittle when

Cordyceps militaris

dry. **STALK:** Continuous with the fertile apical part of the fruiting body, cylindrical, often curved, white, buff, or orange. **SPORES:** Colorless, threadlike, smooth, multiseptate. **HABITAT:** One to several fruiting bodies on underground insect larvae or pupae. **EDIBILITY:** Probably edible; not used medicinally like other species of *Cordyceps*, which are consumed as performance enhancers or in Eastern medicine for general vitality.

Cordyceps militaris is primarily a parasite of the pupae of moths and butterflies, infecting the larvae but often killing the insect only when the insect pupates. **C. sobolifera** is another species of *Cordyceps* that parasitizes cicada insects. **Elaphocordyceps ophioglossoides** (*Cordyceps ophioglossoides*) and **E. capitata** (*Cordyceps capitata*) are both parasitic on the underground truffle *Elaphomyces*.

Hypomyces lactifluorum
Lobster Mushroom

FRUITING BODY: A mass of bright orange, red, to purple-red fungal tissue that roughly resembles the shape of the host mushroom that the fungus parasitizes, often the shape is that of a top

Hypomyces lactifluorum

or inverted pyramid, usually with vertical ridges that mirror the gills of the host; the surface is rough and pimpled from the continuous layer of embedded but protruding red, flask-shaped perithecia that contain the spores of this fungus; the inner flesh is white and the texture is brittle and crisp. **SPORES:** Colorless, spindle shaped with sharp points at both ends, one-septate, ornamented with low warts. **HABITAT:** Parasitic on mushrooms, particularly *Lactarius* and *Russula*. **EDIBILITY:** Although this fungus is edible, caution is advised because the host mushroom may be poisonous. However, reports of poisoning are rare, and at maturity little if any of the original mushroom remains. Poisonings might be attributed to collecting the Lobster Mushroom before it is completely mature and the original mushroom host is not completely digested.

The bright red color of *H. lactifluorum* is distinctive. Other commonly encountered mold species include **H. cervinigenus,** a white or pinkish mold that occurs on *Helvella lacunosa;* **Hypomyces chrysospermus**, which is initially white becoming bright yellow and occurs on boletes; and **H. luteovirens,** a yellow-green mold on *Russula* and *Lactarius* species. These molds are inedible.

Sarcosphaera coronaria

FRUITING BODY: 4 to 12 cm in diameter, initially a hollow subterranean sphere, eventually erupting through the soil surface, at maturity becoming bowl-like with 4 to 10 acute triangular lobes, the fertile inner surface purple or pinkish brown, the outer surface white to cream colored, scurfy with clumps of attached soil and debris, texture fleshy-brittle. **STALK:** Absent or present as a thick base of attachment up to 3 cm long. **SPORES:** Colorless, elliptical with blunt ends, smooth. **HABITAT:** Single, scattered, or clustered under conifers or occasionally oak in the mountains, fruiting in spring. **EDIBILITY:** Unknown; reported as poisonous in some sources and edible in others.

This large and attractive fungus exploits a subterranean habitat to prevent desiccation in its arid mountain home. It is often noticed as purple-lined holes in the ground. *Geopora arenicola* also grows as a hollow sphere (1 to 3 cm broad) below the soil surface and opens when the sides split into several rays. It has a

Sarcosphaera coronaria

yellowish fertile surface and brownish hairy exterior. Like *S. coronaria*, **Peziza violacea** has a purple fertile surface and fruits in spring in the mountains, but it is cup shaped and fruits above ground on burnt soil and charred wood.

Tuber oregonense
Oregon White Truffle

FRUITING BODY: Spherical to irregularly lobed, resembling a small potato, 1 to 5 cm in diameter, whitish when young, developing reddish orange tints as it matures, the peridium (outer rind) cracking in age, the interior solid, white at first, becoming brown, marbled with white veins, odor garliclike. **STALK:** Absent. **SPORES:** Brown, elliptical, reticulate. **HABITAT:** Subterranean, singly to gregarious, mycorrhizal with Douglas-fir, especially trees less than 30 years old; often found between the leaf litter layer and the soil. **EDIBILITY:** Edible and choice.

This truffle is identified by its reddish orange colors of the exterior rind, which develops cracks in age, marbled interior, and reticulate spores born in asci. European truffles such as *T. magnatum* (White Truffle) and *T. melanosporum* (Black Truffle) are culinary delicacies with values as high as $1,000/lb. On the West Coast, *T. oregonense* and *T. gibbosum*, which also goes by

*Tuber
oregonense*

the common name Oregon White Truffle, have been touted as fine alternatives to the expensive European truffles. Because of their subterranean habitat, truffles are difficult to collect. Many species give off an odor attractive to rodents, which eat the fruiting bodies and disseminate the spores. Sometimes truffles can be located by the presence of shallow holes dug by squirrels. In Europe, dogs and pigs are trained to locate truffles. **Elaphomyces granulatus**, which is very common in western U.S. forests, has a thick, warted peridium encrusted with soil. Unlike *Tuber* species, the spore mass is powdery at maturity.

Genea arenaria
Geode Truffle

FRUITING BODY: 1 to 3 cm in diameter, subterranean, irregularly lobed, geodelike, hollow but the cavity irregular in shape from folds of the outer wall, usually with a small inconspicuous opening, the surface pale brown and mealy at first, becoming covered with long brown hairs and ornamented with small conical to pyramidal warts, the inner surface of the chambers covered with small dark warts similar to those on the exterior of the fruiting body, odor mild or garliclike. **STALK:** Absent. **SPORES:** Colorless, elliptical, ornamented with warts. **HABITAT:** Subterranean, my-

Genea arenaria

corrhizal with oaks. **EDIBILITY:** Unknown, but too small and tough to be of interest.

This species is an example of a subterranean (hypogeous) fungus. Similar species include **G. harknessii,** which lacks the brown hairs on the surface of the fruiting body and has nearly spherical spores. It has a dark reddish brown to blackish warty exterior. **Genabea cerebriformis** also lacks external hairs. It has a lumpy, yellowish to pale brown exterior and globose spores. The fruiting bodies of **Gilkeya compacta** (*Genea intermedia*) are reddish to vinaceous, lobed, and warty. It grows with conifers and oaks in the mountains. The crusty exterior of these fungi is ideal for surviving long periods in the dry arid mountains of the western United States.

Sarcoscypha coccinea
Scarlet Cup Fungus

FRUITING BODY: Cup or bowl shaped, 2 to 6 cm in diameter, the fertile inner surface bright scarlet red fading to reddish orange in age, the outer surface whitish and covered with downy hairs, texture tough-fleshy. **STALK:** Absent or, if present, tapering downward, up to 3 cm long, white, and covered with fuzzy hairs. **SPORES:** Colorless, elliptical, smooth. **HABITAT:** Single to gregari-

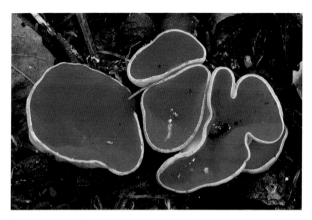

Sarcoscypha coccinea

ous on fallen hardwood sticks and branches, usually fruiting late winter or spring. **EDIBILITY:** Edible but inconsequential.

This beautiful and distinctive red cup fungus often fruits late in the mushroom season. The hairless cup margin distinguishes it from the smaller *Scutellinia scutellata* (p. 398). *Plectania melastoma* has sessile or short-stalked black cups up to 2.5 cm wide with a wrinkled, scurfy, brownish orange exterior, particularly at the margin. It grows in groups on decaying hardwood and conifers. *Plectania nannfeldtii* also has black cups but with relatively long and slender stalks up to 6 cm long. It fruits on decaying conifer wood in the high mountains soon after the snow melts in spring.

Sowerbyella rhenana
(Other name: *Aleuria rhenana*)
FRUITING BODY: Cup shaped, 1 to 2 cm in diameter, the inner fertile surface bright orange to yellowish orange, the outer surface orange and covered with white, fuzzy hairs when young, becoming smooth in age, texture brittle. **STALK:** 1 to 3 cm long and up to 5 mm or more in diameter, white to pale orange, fuzzy to smooth. **SPORES:** Colorless, elliptical, reticulate with connected

Sowerbyella rhenana

warts. **HABITAT:** Gregarious to clustered on soil, mycorrhizal with conifers and possibly other trees. **EDIBILITY:** Unknown.

This orange cup fungus is very similar to the more common **Aleuria aurantia** (p. 406), which is larger, lacks the distinct stalk of *S. rhenana*, and is usually found under hardwoods. **Caloscypha fulgens** (p. 405) is also an orange cup fungus, but it bruises bluish green, a reaction absent in *S. rhenana.* Other common orange cup and disklike fungi in the West include **Anthracobia melaloma** (p. 389), which has tiny brown hairs around the disk and grows on burnt soil, and **Octospora** species, which have small cups typically just a few millimeters broad found on or near moss. Species of *Anthracobia* and *Octospora* lack stalks and the distinctive spore ornamentation of *S. rhenana.*

Geopyxis carbonaria

FRUITING BODY: 5 to 15 mm in diameter, deeply cup shaped, inner fertile surface orange-brown to reddish brown and smooth, external surface brown and smooth, the margin of the cup paler and delicately toothed, flesh thin and pliable to brittle. **STALK:** Usually present, short, tapered, up to 1 cm long, whitish to light brown. **SPORES:** Colorless, elliptical, smooth, without oil droplets. **HABITAT:** Scattered or gregarious on burnt ground,

Geopyxis carbonaria (brown cups) and *Anthracobia melaloma* (orange disks)

fruiting in spring. **EDIBILITY:** Unknown, too small to be of interest.

The brown fertile surface of the cup, finely toothed margin, and presence of a stalk are the principal field marks of this common cup fungus of burnt soil. **Geopyxis vulcanalis** is also deeply cup shaped, but it becomes saucer shaped in age. The fertile surface is yellowish or pale orange, whereas the outer surface is paler. It grows on soil in coniferous forests, often among mosses. **Tarzetta catinus** has a cream-colored to yellowish brown cup 1 to 4 cm broad with a toothed margin and granular outer surface, and usually a well defined stalk. It grows on the ground in woods. The cup of **T. cupularis** is generally less than 15 mm broad and gray or cream colored. It is found on soil, among moss, and in burn sites. As a group, the major distinguishing feature of *Tarzetta* species is the presence of two oil drops in the spores.

Trichoglossum hirsutum
Velvety-black Earth Tongue

FRUITING BODY: Club, lance, or spoon shaped, fertile cap 1 to 2 cm long and 2 to 8 mm wide, laterally compressed and often with longitudinal folds or creases, resembling a partly deflated balloon, black, dry, tough, velvety from the numerous projecting setae (stiff hairlike hyphae). **STALK:** 3 to 6 cm long, 1 to 4 mm wide, equal, smoothly transitioning into the cap, frequently twisted and curved, black, velvety. **SPORES:** Brown, needle shaped, 13 to 16 septate, smooth. **HABITAT:** Single, scattered, or in tufts in moss, on rotten wood, or under a variety of trees. **EDIBILITY:** Inedible.

As a group, earth tongues are identified by their upright, clublike fruiting bodies. Because of their dark coloration, they blend into the forest floor and are easily overlooked. ***Geoglossum fallax*** is another dark earth tongue encountered in mixed hardwood-coniferous forests. It has an overall dark brown color and a minutely roughened, not hairy, exterior, unlike the velvety covering present in *T. hirsutum*. They sometimes grow together. The Green Earth Tongue, ***Microglossum viride,*** is similar but green. The cap is often flattened and usually furrowed, and the stalk is scurfy.

Trichoglossum hirsutum

Leotia lubrica
Jelly Babies

CAP: 1 to 3 cm broad, rounded to convex, often lobed, upper surface (fertile surface) smooth, slightly wrinkled, or convoluted, pale green to yellowish green to yellow, viscid when moist, the margin inrolled and wavy, the underside (sterile surface) yellowish or pale green, flesh gelatinous. **STALK:** 2 to 8 cm long, 0.5 to 1 cm wide, more-or-less equal, viscid when moist, yellowish or colored like the cap, smooth or roughened, hollow or filled with a gel. **SPORES:** White, spindle shaped, curved, septate, smooth, produced in asci. **HABITAT:** Single, gregarious, or in groups on ground under hardwoods and conifers, sometimes on very rotten wood. **EDIBILITY:** Edible but inconsequential.

The pale green or yellowish fertile surface of the cap, yellow or orange stalk, and gelatinous texture are a distinctive set of characteristics. The species name has the same root as "lubrication" and refers to the cap's slippery feel. The cap will lose color when dry. *Leotia viscosa*, called Chicken Lips, is similar, but it has an olive to dark green, viscid cap and a yellow or pale tan viscid stalk. It is also found on the ground or on rotten wood in forests.

Leotia lubrica

Morchella rufobrunnea

Morchella rufobrunnea
Morel

CAP: 6 to 12 cm high, 2 to 6 cm in diameter, conical with deep concave pits bounded on all sides with intersecting ridges, the pits and ridges usually arranged vertically, overall tan to yellow-brown but the ridges paler when young, the cap attached for its length to the stalk, texture firm-fleshy to rubbery, bruising reddish brown, odor and taste mild. **STALK:** 3 to 10 cm tall, 1 to 3 cm wide, cylindrical with longitudinal folds, hollow, white to buff, often with brownish stains, smooth or covered with mealy particles, especially in age. **SPORES:** Pale orange, elliptical, smooth. **HABITAT:** Single to clustered on disturbed ground and in landscaped areas, typically fruiting in spring. **EDIBILITY:** Edible and choice.

Morchella rufobrunnea is the morel encountered in landscaped areas in the West. It sometimes fruits in large numbers the year after wood chips are spread on the ground as mulch. The pale tan pits and ridges, which are elongated vertically, have a tendency to bruise reddish brown. ***Morchella esculenta*** represents several unnamed species that occur in a variety of habitats, including woods, stream edges, and under deciduous trees. None is likely the true *M. esculenta* first described in Europe, but because morel taxonomy is extremely confused, a more accurate name is not available. Other species include **M. elata,** the Black Morel (p. 2), which is commonly collected in montane forests, especially a year after a forest fire. The taxonomy of that species is also in flux. Only the top half of the cap of **Mitrophora semilibera** (*Morchella semilibera*) is attached to the stalk. All of these species should be compared with species of *Gyromitra,* a potentially toxic mushroom that has brainlike folds.

Verpa conica
Smooth Thimble Cap

CAP: 1 to 4 cm tall and broad, bell shaped or conical, the lower edge free and skirtlike, smooth, wavy, occasionally with a few faint veins or furrows, tan, olive brown, or dark brown, the edge sometimes whitish, the underside whitish, finely hairy, the cap attached to the stalk only at the very top, odor and taste mild. **STALK:** 2 to 12 cm tall, 1 to 3 cm wide, cylindrical or slightly enlarged downward, white to cream colored, smooth to scurfy, sometimes marked with scaly concentric rings. **SPORES:** Colorless, elliptical, smooth. **HABITAT:** Single to gregarious in litter under both conifers and hardwoods, fruiting in spring. **EDIBILITY:** Edible but of lesser quality than the true morels.

Verpa conica

Verpa conica is characterized by a brown conical cap attached only at the tip of the whitish stalk and spring growth habit. **Verpa bohemica** is similar but has a netlike wrinkled cap, causing confusion with morels. True morels (*Morchella* species), however, have pitted and ridged caps typically attached to the stalk for their entire length. One exception is **Mitrophora semilibera** (*Morchella semilibera*); only the top half of its cap is attached to the stalk.

Gyromitra esculenta
False Morel

CAP: 3 to 10 cm tall and broad, spherical to elliptical, surface folded, convoluted and brainlike with several lobes, yellow-brown, reddish brown, or dark brown, undersurface paler than the top surface, cap attached to the stalk in several places, otherwise closely adhering to the stalk, flesh waxy-rubbery to brittle, odor and taste mild. **STALK:** 3 to 6 cm long, 1 to 5 cm wide, cylindrical, often with longitudinal furrows, whitish, sometimes tinged reddish orange, smooth. **SPORES:** White to pale yellow, elliptical, smooth, with two oil droplets. **HABITAT:** Single, scattered, or gregarious on the ground in mixed woods, especially

Gyromitra esculenta

common under conifers in high mountains in spring. **EDIBILITY:** Toxic due to the presence of monomethylhydrozine. The toxin is highly volatile and can be eliminated by thorough cooking, but caution is advised. Although apparently few or no confirmed poisonings have been reported in the western United States, many poisonings have been reported in the eastern United States and Europe.

The distinguishing features of the False Morel are the brain-like, reddish brown cap, and growth in spring. ***Gyromitra gigas*** (*G. montana*) is similar, but it has a stocky stature, and its yellow-brown, convoluted cap lacks lobes (p. 425). The reddish brown cap of **G. infula** (p. 41) is saddle shaped and can be confused with species of *Helvella*.

Helvella acetabulum

CAP: Up to 8 cm broad, cup shaped, the margin entire or sometimes split, the fertile interior surface brown and smooth, the exterior sterile surface brown blending into light brown and finally white or creamy at the base, darkening to almost black in

Helvella acetabulum

age, conspicuously ribbed, the raised ribs branching and extending toward the margin of the cup, flesh thin and brittle, odor and taste mild. **STALK:** Up to 3 cm long and 3 cm wide, whitish, ribbed. **SPORES:** Colorless, elliptical, and smooth with a single oil droplet. **HABITAT:** Single or scattered on ground in leaf litter under a variety of trees, especially oak; fruiting in late winter and spring. **EDIBILITY:** Unknown.

Conspicuous whitish ribs on the stalk extend up the exterior of this brown saddle fungus. Other cup-shaped saddle fungi include *H. leucomelaena*, which has a brown interior, a smooth brownish exterior, and a short white stalk. It grows on the ground under pines and other conifers. *Helvella queletii* is cup shaped or almost flat in age; the cup is brown to almost black with a smooth or scurfy exterior. The ribs of the white stalk end at the base of the cup.

Helvella lacunosa
Black Elfin Saddle

CAP: 2 to 8 cm high, 2 to 5 cm broad, irregularly lobed, convoluted, or saddle shaped, fertile (upper surface) black or dark gray and smooth, underside (sterile surface) gray and smooth or ribbed, cap attached to the stalk in several places, texture brittle. **STALK:** 3 to 15 cm long, 1 to 3 cm wide, equal or tapering slightly toward the apex, smooth, longitudinally ribbed, forming pockets (lacunose), whitish, gray, or dark gray, darkest at the apex, internally chambered. **SPORES:** Colorless, broadly elliptical, smooth, with a single oil droplet. **HABITAT:** Single, scattered, or gregarious on soil under a variety of trees. **EDIBILITY:** Edible, but there have been reports of stomach upsets. A small amount of the toxin monomethylhydrozine may be present, so the mushroom should be thoroughly cooked.

This is the most common of the *Helvella* species found in the West. It is highly variable, but the conspicuously ribbed stalk and convoluted black cap identify it. Other saddle fungi with fluted stalks include *H. maculata*, which has a brownish mottled cap with a hairy undersurface, and *H. crispa*, which has a whitish cap. The primary features of *H. elastica* are the grayish brown color of the fertile surface of the saddle-shaped cap, smooth whitish undersurface, and creamy smooth stalk.

Helvella lacunosa

Bulgaria inquinans
Black Jelly Drops
FRUITING BODY: 1 to 4 cm high and broad, at first roughly globose, becoming slightly depressed, broadly convex, or flattened in age, black or dark brown, shiny when moist, duller when dry,

Bulgaria inquinans

outer surfaces brown and scurfy, flesh tough and rubbery when young, becoming gelatinous. **STALK:** Absent or funnel shaped, brown to black, mostly buried in the host substrate. **SPORES:** Brown, elliptical to kidney shaped, smooth. **HABITAT:** Single, scattered, or in swarms on dead hardwood logs and branches, especially common on oak. **EDIBILITY:** Unknown.

This distinctive fungus, which looks like licorice candy, is easily confused with some jelly fungi because of its rubbery or gelatinous texture. Jelly fungi, however, produce spores on basidia, whereas this fungus produces spores in saclike asci. Young specimens resemble small black cups but become more free-form as they age. The jelly fungus *Exidia glandulosa* (p. 306) has black, convoluted, gelatinous fruiting bodies that are smooth or dotted with small warts. It often fruits in large masses, concealing individual fruiting bodies. *Ascocoryne sarcoides* (p. 400) is a relative of *Bulgaria inquinans* and also grows on wood, but it is purplish in color.

Scutellinia scutellata
Eyelash Cup

FRUITING BODY: 4 to 12 mm in diameter, 1 to 4 mm high, saucer to disk shaped, inner surface bright red or orange, occasionally fading in age to yellow or tan, smooth, the margin decorated

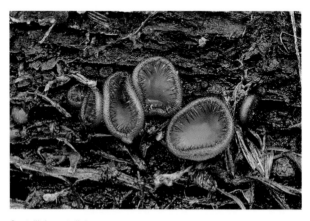

Scutellinia scutellata

with dark brown to black stiff hairs up to 2 mm long, the under-surface similar in color to the upper side but appearing darker due to the presence of dark hairs, texture fleshy. **STALK:** Absent. **SPORES:** Colorless, elliptical, smooth or minutely warted, with one to several oil droplets. **HABITAT:** Scattered to crowded on well-decayed wood or bark, especially oak, fruiting in spring but sometimes found at other times of the year. **EDIBILITY:** Too small to be of interest.

The bright red color of the fruiting body, blackish hairs that line the margin of the disk, and growth on wood are the primary characteristics of the Eyelash Cup. A host of other red *Scutellinia* species can be differentiated by microscopic characteristics of hairs and spores. For example, **S. umbrarum** has distinctly warted spores. It grows on damp soil and on rotted wood. **Lachnellula arida** (p. 401) is similar, but it has a yellow cup lined with short brown hairs. It grows on fallen conifer twigs and branches in the mountains in spring soon after the snow melts.

Peziza repanda
FRUITING BODY: 2 to 12 cm in diameter, bowl or saucer shaped, becoming wavy and nearly flat, orange-brown, fawn-brown, or dark chestnut brown, the upper surface smooth to wrinkled, the

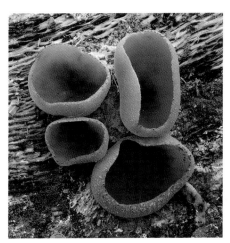

Peziza repanda

margin even to scalloped, often splitting as the fruiting body flattens, the underside white to tan, smooth, scurfy, or fuzzy, texture brittle, odor and taste mild. **STALK:** Short to absent. **SPORES:** Colorless, elliptical, smooth, lacking oil droplets. **HABITAT:** Scattered to clustered, usually on rotted logs, wood chips, decaying roots, and such, in woods and in landscaped areas. **EDIBILITY:** Unknown.

A large number of *Peziza* species in the West are difficult to identify. *Peziza repanda* is identified by its large size, tendency to flatten in age, and occurrence on rotten wood. The woodland **P. arvernensis** (*P. sylvestris*) has a yellowish brown fruiting body and spores ornamented with fine warts; **P. vesiculosa** has a brown fruiting body 2 to 5 cm in diameter that remains cup shaped, even in age, has smooth spores, and fruits on horse manure or compost piles; **P. domiciliana** is noteworthy because it fruits on wet rugs in houses, among other substrates; and **P. violacea** is a small purplish brown species that occurs on burnt soil and charred wood.

Ascocoryne sarcoides

FRUITING BODY: 5 to 10 mm in diameter, spherical to turban or cup shaped at first, often coalescing into an irregular mass, each cup flattened or becoming concave in age, shiny, violet, purple, or reddish purple, smooth, flesh gelatinous, odor and taste mild. **STALK:** Absent or, if present, a short extension of the cap. **SPORES:** Colorless, elliptical to spindle shaped, smooth, one to three septate. **HABITAT:** Gregarious or clustered on hardwood logs and stumps, especially on the cut ends of logs. **EDIBILITY:** Unknown but too small to be of value.

The violet to purple coloration of the clustered cups and wood habitat are the key features of *A. sarcoides*. **Bulgaria inquinans** (p. 397) is a close relative, but it is black in color. Both of these species might be confused with jelly fungi because of their gelatinous texture, but they have spores borne in asci instead of on basidia like jelly fungi. The jelly fungi **Auricularia auricula** (p. 305) and **Tremella foliacea** are brown in color, larger than *A. sarcoides*, and have earlike or leaflike lobes. Another jelly fungus, **Exidia glandulosa** (p. 306), has black, convoluted, gelatinous fruiting bodies.

Ascocoryne sarcoides

Lachnellula arida

FRUITING BODY: Round, flat to cuplike, bright yellow to creamy yellow, 2 to 6 mm in diameter, usually sessile, the cup exterior covered with short brown hairs. In dry weather, the cup shrivels and the fertile yellowish surface may not be readily visible; in

Lachnellula arida

wet weather, the fruiting body rehydrates and revives. **SPORES:** One celled, elliptical, smooth. **HABITAT:** Gregarious on fallen conifer twigs and branches, common in the mountains in spring soon after the snow melts but also present in summer after rain. **EDIBILITY:** Unknown but too small to be of value.

Several species of *Lachnellula* occur on conifer wood and debris. In general, they all have yellowish fertile surfaces, but they differ based on the color of the exterior of the cups, the size and shape of the spores, and other microscopic characteristics. ***Bisporella citrina*** (below) is another common, small, yellow disk- or saucer-shaped fungus. It grows on the surfaces of hardwood branches that have lost their bark. Unlike *L. arida*, the exterior surface of *B. citrina* lacks hairs. The spores of *B. citrina* are two celled, with conspicuous oil droplets. See *B. citrina* (below) for other yellow disklike fungi.

Bisporella citrina

FRUITING BODY: Saucer to disk shaped, less than 3 mm across, deep yellow, exterior smooth, texture tough-gelatinous. **STALK:** Very short and thick, gradually narrowing from the disk, pale yellow. **SPORES:** Colorless, smooth, elliptical to nearly rod shaped, one celled when young, becoming two celled in age with conspicuous oil droplets. **HABITAT:** Gregarious, usually in

Bisporella citrina

swarms that may become confluent masses, on dead barkless hardwood. **EDIBILITY:** Much too small to be considered.

Bisporella sulfurina is very similar (small, yellow, and disk shaped), but it fruits in clusters rather than swarms. Other similar species include *B. pallescens*, which has small whitish cups. It fruits on dead hardwoods. Sessile yellow or orange cup fungi include *Anthracobia melaloma* (p. 389), which has a minutely hairy cap margin and preference for burnt soil, and *Octospora* species, which have small cups typically just a few millimeters broad found on or near moss. *Guepiniopsis alpina* (p. 307) is a larger, yellow, disk-shaped jelly fungus that occurs on dead conifer wood in the high mountains, especially after the snow melts in spring. *Dacrymyces capitatus*, another yellow jelly fungus, is cushion shaped and might be mistaken for *B. citrina*. However, both *H. alpinus* and *D. capitatus* produce their spores on basidia rather than in asci.

Otidea alutacea

FRUITING BODY: Cup 3 to 6 cm high and 2 to 4 cm wide, more-or-less erect, asymmetrical, one side usually split to the base and opened, tapering toward the base, truncate at the apex, the margin often wavy, the inner surface pale brown or grayish brown,

Otidea alutacea

the outer surface brown and mealy to almost smooth, flesh brittle, odor and taste mild. **STALK:** A short point of attachment or absent, the base covered with fine white hairs. **SPORES:** Colorless, elliptical, smooth with two oil drops. **HABITAT:** Scattered or more often in clusters on ground in mycorrhizal association with conifers and hardwoods. **EDIBILITY:** Unknown.

The brownish, truncated, cuplike fruiting bodies of *O. alutacea* commonly grow in clusters. **Otidea onotica** is similar, but it is usually elongated and split to the base on one side, giving it a rabbit-ear–like appearance because it stands semierect. The inner surface is ochre, yellow, or orange and often flushed with pink, whereas the outer is similarly colored but lacks the pinkish tints. It goes by the common name Donkey Ears. **Guepinia helvelloides** (*Phlogiotis helvelloides*) is a somewhat similar pinkish orange, tonguelike jelly fungus. It has spores borne on basidia; in *Otidea* species, spores are borne in sacs.

Discina perlata
(Other names: *Discina ancilis, Gyromitra perlata*)

FRUITING BODY: 2 to 10 cm broad, cup or saucer shaped becoming flattened and often centrally depressed, the margin wavy and curved down in age, the upper fertile surface cinnamon brown to yellow-brown, nearly smooth to wrinkled or convoluted, the outer surface whitish becoming brownish, smooth to velvety, flesh white, brittle, and thick (up to 3 mm), odor and taste mild. **STALK:** Absent or, if present, up to 1 cm or more long, tapered toward the base, cream colored to brown with conspicuous folds or ridges. **SPORES:** Colorless, elliptical to spindle shaped with a sharply pointed appendage at each end, smooth when young becoming minutely warted at maturity, with oil droplets. **HABITAT:** Single, gregarious, or in small clusters on ground or on very rotten wood in montane forests after snow melt. **EDIBILITY:** Edible, but it should be cooked thoroughly.

Discina perlata is a conspicuous snowbank fungus that often begins development under the snow. The brown cuplike fruiting bodies become flattened and often wrinkled to various degrees. Note the pollen grain on the fruiting body in the photograph, confirming the springtime habit of this species. **Peziza repanda** (p. 399) might be confused with *D. perlata*, but *Peziza* has thin flesh, never has a stalk, and grows on wood or wood chips. **Dis-**

Discina perlata

ciotis venosa, which is also brown and cup to saucer shaped and grows on the ground, has a veined or wrinkled upper fertile surface. Unlike *P. repanda,* the asci of *Disciotis venosa* are inamyloid, and unlike *Discina perlata, Disciotis venosa* has spores that lack oil droplets.

Caloscypha fulgens
FRUITING BODY: 1 to 6 cm in diameter, cup to saucer shaped or nearly flat in age, sometimes split on one side, inner fertile surface bright orange and smooth, outer surface yellow to orange, often with an olive green marginal band, bruising bluish green or becoming that color in age, flesh thin and brittle. **STALK:** Absent or present as a short white base. **SPORES:** Colorless, globose, smooth. **HABITAT:** Scattered, gregarious, or clustered on soil or in duff under conifers, often near melting snow in the mountains. **EDIBILITY:** Unknown.

This common snowbank fungus is easily identified by the orange color of the cups and the bluish green staining reaction when handled. *Caloscypha fulgens* is apparently a pathogen of seeds and plant roots during cool temperatures when seeds and plants are dormant. At higher temperatures, when plants and

Caloscypha fulgens

seeds are active, the fungus is no longer an effective parasite. It is frequently encountered near winter caches of conifer cones made by squirrels. ***Aleuria aurantia*** (below) is similar, but it lacks the staining reaction and has ornamented spores. It also typically fruits in fall, not spring, an infallible field mark of *C. fulgens.*

Aleuria aurantia
Orange Peel

FRUITING BODY: 2 to 10 cm in diameter, cup shaped, but becoming irregular from pressure of other adjoining fruiting bodies, often flattening and splitting in age, inner fertile surface bright orange, outer surface pale orange or yellow, covered with white, downy hairs when young becoming smooth in age, flesh thin and brittle, odor mild. **STALK:** Absent or rudimentary. **SPORES:** Colorless, elliptical, reticulate (ornamentation resembling a honeycomb), produced in asci on the inner surface of the cup. **HABITAT:** Gregarious to clustered, on ground in woods, along paths, in gardens, disturbed areas, and wood chips. **EDIBILITY:** Edible; sometimes used to add color to a salad or dessert.

This cup fungus is readily recognized by its bright orange color. Not surprisingly, the fruiting body resembles a discarded

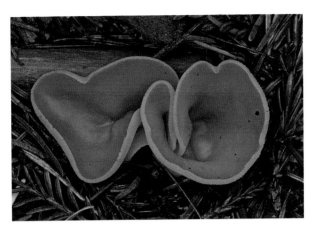

Aleuria aurantia

orange peel. Similar species include **Sowerbyella rhenana** (p. 387), a smaller bright orange cup fungus with a distinct stalk; the snowbank cup fungus **Caloscypha fulgens** (p. 405), which is bright orange with bluish green stains, a color absent in *A. aurantia;* species of **Peziza** (p. 399), which are often brownish; and species of **Otidea** (p. 403), which are asymmetrical and generally more erect. **Sarcoscypha coccinea** (p. 386) is a bright red cup fungus that occurs on wood.

SLIME MOLDS

Arcyria denudata

FRUITING BODY: Cylindrical, erect, up to 6 mm tall, bright red or pink, containing persistent pink threads. **STALK:** Up to 2 mm long, reddish. **SPORES:** Bright red to reddish brown in mass, globose, minutely warty. **HABITAT:** Gregarious to crowded on decaying wood and plant debris; the plasmodium is white. **EDIBILITY:** Inedible.

Like other slime molds, *A. denudata* creeps amoeba-like inside rotted wood in search of bacteria, protozoa, and other microscopic life that it engulfs. When it reproduces, it crawls to the surface of wood to produce a mass of fruiting bodies. After spore release, the persistent pink threads in the empty sporangia (spore cases) look like cotton candy. **Stemonitis fusca** has dark reddish brown, cylindrical fruiting bodies up to 20 mm tall with shiny black stalks. The plasmodium is white. ***Ceratiomyxa fru-***

Arcyria denudata

ticulosa produces white, erect, simple or branched columns up to 1 cm tall in patches up to 10 cm broad. It differs from other slime molds in that the spores are produced on the outside of the fruiting bodies. **Cribraria cancellata** (*Dictydium cancellatum*) has reddish brown, globose sporangia less than 1 mm in diameter on stalks attached to wood. After spore release the walls of the sporangia remain as longitudinal ribs connected by transverse bands, forming a basketlike net. The spores are reddish to purple, and the plasmodium is purple-black.

Fuligo septica
Dog Vomit Slime Mold

FRUITING BODY: Initially spongelike with rounded margins, up to 20 cm or more broad and 1 to 3 cm thick, the outer surface turning white or pinkish white and finally gray as the spores within mature and turn black, the interior marbled with white or yellowish threads. **STALK:** Absent. **SPORES:** Pinkish brown to dull black, spherical, minutely spiny. **HABITAT:** On stumps, logs, mulches, decaying plant material, lawns, agricultural fields, and so on; the plasmodium is bright yellow. **EDIBILITY:** Inedible.

Fuligo septica, which is not a true fungus, passes most of its life as a gigantic, yellow, cellular mass called a plasmodium 20 cm or more in diameter (inset). The protoplasm within the plas-

Fuligo septica plasmodium (inset) and mature fruiting body.

modium pulsates in an ebb and flow movement, allowing the creature to slowly move amoeba-like in and on rotting vegetation in search of bacteria, fungi, protozoa and other microscopic life, which it engulfs. To sporulate, it moves to the surface and turns itself into a mass of dry spores that are disseminated by wind and insects. The irregular shape of the fruiting body is determined in part by the shape of the plasmodium as it becomes spores. The large size, bright color, ability to move, and its sudden appearance in lawns and gardens has prompted declarations of alien invasions.

Lycogala epidendrum

FRUITING BODY: Globose or cushion shaped (puffball-like), bright pink when young becoming pinkish gray, olive, or black, up to 15 mm broad, the surface roughened. **STALK:** Absent. **SPORES:** Globose, at first pink in mass and appearing like toothpaste, becoming pale yellowish brown and powdery in maturity. **HABITAT:** Scattered or in crowded groups on decaying wood; the red plasmodium, which moves amoeba-like, worms it way among the fibers of wet, well-rotted wood, devouring its prey of microscopic animals, bacteria, and fungi. **EDIBILITY:** Much too small to be of interest.

Lycogala epidendrum

Other common slime molds include **Physarum nutans**, which produces large colonies of gray, stalked, globose or flattened fruiting bodies up to 1 mm broad. The stalk is brown and tapered toward the top. **Physarum cinereum** is similar, but it produces sessile sporangia (spore cases) up to 0.5 mm broad. It fruits on woody debris and living plants, including grass blades, in crowded masses. **Hemitrichia calyculata** produces red and finally yellowish gobletlike fruiting bodies that appear to contain a foamy beverage. The plasmodium is pink to red. **Trichia varia** produces yellow globose fruiting bodies that are sessile or occasionally on short stalks. The spores are yellow. Like most slime molds, it fruits on decaying wood.

Sketching Mushrooms

Why sketch or paint pictures of mushrooms when photography is quicker and easier, and a macro lens captures exceptional detail? Mushroomers sketch in order to see and to remember. Drawing captures the critical features of a thing, its purity and individuality, what sets it apart from other things (pp. 424–425). With photography, there is the risk of paying more attention to technology than to what you see. To paraphrase John Norwich in his book *Venice in Old Photographs*, "In an ideal world it would surely be illegal to photograph anything without having looked at it for at least five minutes." For John Ruskin, even five minutes would not have been sufficient. "Don't look at [things]," he told his students, "watch them." Drawing requires a person to stop and look intently, to attend to detail and nuance that would otherwise be missed.

If you are unsure of your drawing ability, start sketching, and don't show the early drawings to anyone. Experiment with different media—watercolor, black ink by itself, black ink and watercolor wash, and colored pencils. See which you like best. As the quality of your artwork improves (and with practice, it will), consider buying a better grade of paper. A standard size is helpful for storage but will prove too large for some fungi and too small for others. You will probably want two paper sizes, a pocket-size 4 × 6 inch pad for small mushrooms and a larger 9 × 12 inch sketchbook for medium and large fungi. Over time the sketches will become a collection and require a filing and storage system.

Paint fresh specimens when possible. Within a few hours, cap color can fade, white gills can turn brown, and brightly colored waxy caps may blacken. Older specimens have their appeal to the artist, with interesting cracks, bruises, or color changes. You will appreciate dirt on a light-colored cap, the yellow stains on *Agaricus xanthodermus*, an inky cap whose gills are turning into ink, a cracked puffball, and a bluing bolete. These "blemishes" add interest to a sketch. Favorite mushroom models include anything with a cup, ring, warts, raggedy cap, or fibrillose stalk; mushrooms with a distinctive color or stately shape; large fungi as long as they fit on the page; and species with stains that are a contrasting hue. Least preferred models might be pure white or translucent fungi; tiny mushrooms that get lost on the page; the Beefsteak Fungus (*Fistulina hepatica*), whose "blood" stains everything; and buggy mushrooms whose frantic inhabitants scamper across the page.

Sketches often look better a few days after their completion. By this time, the model has been discarded and minor distortions in color or detail are no longer recognized. Judge the worth of a completed sketch after the image of the original model has faded.

Artistic Spore Prints

With their symmetry and color, spore prints can be very attractive. You can make single prints from large-capped mushrooms, or you can combine several smaller mushroom caps of the same or different species to create a collage. Obtain good-quality paper or a canvas for your artistic spore prints, and select species that are heavy spore droppers. Select light-colored paper for dark-spored species and dark-colored paper for white-spored

A collection of spore prints on a framed canvas.

Amanita velosa painted with puffball spores.

species. Some particularly good spore droppers are the Honey Mushroom (*Armillaria mellea*) and Fly Agaric (*Amanita muscaria*), both with white spores; *Volvopluteus gloiocephalus*, with pinkish brown spores; and the Meadow Mushroom (*Agaricus campestris*), with chocolate brown spores. After you have made an artistic spore print, spray it with a fixative (at a distance of 6 to 8 in. to avoid disturbing the spore deposit) to preserve the print.

Painting with Puffball Spores

The spore mass of puffballs turns dark and powdery with age. Mixed with water, it makes a suitable paint medium. Other suitable spore masses for painting are found in *Scleroderma cepa* and *Agaricus deserticola*. Mix the powder with a small amount of water, stir, and paint. Afterward, spray the painting with a fixative to ensure durability. The mixture might be improved for painting if blended or boiled, but good results can be obtained using the stirred raw mixture.

Mycophilately

Maggie Rogers, the doyenne of mushroom stamps, recounts the pleasures of collecting, cataloging, and storing very small things. Regarding postage stamps with mushroom illustrations, she asks rhetorically, "How can one be so captivated by tiny bits of colored paper?" Her answer refers to similarities between collecting stamps from dealers' catalogs and stamp shops, and hunting small mushrooms in the forest. When collecting mushroom stamps, you don't have to worry about a lack of moisture, an early frost, or foraging in a heavy downpour. Unlike forest fungi, mushroom stamps can be kept indefinitely in their original form when stored in glassine sleeves.

An article in a mushroom society newsletter listed mushroom stamps by country of origin, species, and genus. At the time, the two genera depicted most often on postage stamps worldwide were *Amanita* and *Boletus*, and the two most popular species were the King Bolete (*Boletus edulis*), and the yellow-orange chanterelles (*Cantharellus* species) (as described in *The Puffball*, the newsletter of the Willamette Valley Mushroom Society [August/September 1998, 16(4)].

Never rely on mushroom stamps for identifying fungi. With their small size and often dubious illustrations, they are as bad

A collection of international stamps featuring mushrooms.

as or worse than coffee table books for identification. Many drawings are the work of freelance artists who lack mushroom eyes and don't know the distinctive features that set one species apart from another. Discovering instances when mushroom drawings are unrealistic or mislabeled is part of this absorbing hobby.

If you are willing to pay a premium, you can create your own legal U.S. postage stamps with mushroom images or any other images. Check the Photo.Stamps.com website for details.

Photography

The camera is an important tool in mushrooming. The website of the North American Mycological Association lists three goals for mycophotography: teaching, recording and vouchering specimens, and showing the beauty of mushrooms. Mushroom clubs sponsor photo exhibits and slide shows and place outstanding images on their websites. As part of documentation, mushroom photography can assist in learning mushroom names. As a hobby, it can become an art project. Documentary pictures are quick, easy, and inexpensive. One can use a small digital point-and-shoot camera with built-in flash and macro capability to record salient features of mushrooms you want to identify. For documentation, be sure to place several mushrooms in the photo with different orientations showing the cap from above, another seen from the side showing details of the stalk, and a third illustrating the gill structure. Be sure to show whether the mushrooms grow in a cluster, in loose groups, or in isolation. If the stalk stains or bruises a different color, or cut gills ooze latex, show or stage this on your model. If you want, you can include a penny or a ruler to illustrate comparative size, although a twig or leaf in the background often provides scale. The goal is to portray critical information in visual form. Carry a notebook to record picture number, location, and other information such as habitat, and note nearby tree species. Be sure your notes are correctly attributed to your specimens. Collected mushrooms can be placed in paper bags or folded wax paper with an identifying number matching your photograph.

You can also take the occasional art photograph with a point-and-shoot camera. Digital editing afterward, with Photoshop or similar software, can compensate somewhat for the lack of equipment in the field, but the results aren't likely to be as professional as with a digital single-lens reflex (DSLR) camera with interchangeable lenses. There is no limit to how much time and money you can invest in serious mycophotography. You will understand what is involved when you see Taylor Lockwood's videos or Roger Phillips's book illustrations. Although a single camera with a macro feature will do a good job for most situations, serious photographers will invest in additional lenses, lighting equipment, tripod, filters, and more. It may take 30 minutes or more to set up the scene properly, clean dirt from your model with a mushroom brush, remove distracting grass and twigs from the foreground, adjust your mushroom model to the proper angle, and carefully arrange strobe lights and reflectors. This is a technical production, not a hit-and-run photo shoot in which you quickly take multiple images to inspect at home in the hopes that some turn out well. With art photogra-

Mycena adscendens, a small photogenic mushroom.

phy, you look at the images immediately and reshoot if they are not up to your standards.

Not all great mushroom photos require large, showy specimens. A carpet of small mushrooms covering the ground, an elegant fairy ring, or an army of small bright yellow fungi marching up the trunk of a dead tree can make a striking picture. The same is true of a huge fruiting of off-white oyster mushrooms, a close-up of a Honey Mushroom cluster, or an inky cap turning to ink. Before going out into the field with your camera, it is useful to know species that are likely to be fruiting.

Mushroom Dyes

Until the nineteenth century, natural materials such as plants, berries, minerals and earth pigments, tree bark, lichen, and even insects (cochineal) or mollusks (squid) were the source of most dyes. Today, synthetic dyes dominate commercial manufacture but have not eliminated interest in natural colorings among weavers and other craftspeople. Making dye from mushrooms is empirical in that it encourages experimentation, often with unpredicted results. When you boil mushrooms, the resulting hue depends on the type of water used (although you can standardize this by always using distilled water), mushroom species, mushroom condition (young or mature), quantity (ratio of mushroom pieces to water); mushroom parts (cap or stalk), the mordant added, and even the type of pot in which the mixture is boiled—for example, cast iron, aluminum, or copper. Mordants are metallic salts used to aid in the setting of a dye on fiber to make it lightfast and colorfast. Different mordants produce different colors. In their book *Mushrooms for Dyes, Paper, Pigments and Myco-Stix* (2007), Miriam Rice and Dorothy Beebee provide detailed instructions on the selection and use of mordants and the likely color outcomes produced by different mushroom species. All primaries and most intermediate colors can be created using some combination of mushroom species and mordant. The following table shows some of the colors found with different mushroom/mordant combinations. Rice later discovered that the residue left in the dyepot after boiling could be used to make paper. Some of the best mushrooms for

Wool cap knitted with yarn dyed with mushrooms. Pale yellow,
Hypholoma fasciculare; warm brown, *Phaeolus schweinitzii*; cigar brown,
Pisolithus arhizus; blue and gray, *Omphalotus olivascens*; green, *Scutiger
pes-caprae*; pink, *Cortinarius phoeniceus* var. *occidentalis*.

papermaking are polypores such as Artist's Conk (*Ganoderma
applanatum*), Turkey Tail (*Trametes versicolor*), and Red-belted
Conk (*Fomitopsis pinicola*). Their 2007 book documents the use
of dyepot residues to make bowls and sculpture, to form crayons
for drawing and painting, and to make natural inks for wood-
block printing.

Another excellent book on mushroom dyes is *The Rainbow
Beneath My Feet: A Mushroom Dyer's Field Guide* (2001), by Ar-
leen and Alan Bessette. For more on papermaking, see the website of the Western Montana Mycological Association (www
.fungaljungal.org/orgfiles/social/papermake.html).

Dye Colors Produced by Different Mushroom/Mordant Combinations

Color	Mushroom species	Mordant
Black	*Phaeolus schweinitzii*	Iron
Blue	*Sarcodon imbricatus*	Alum
Brown	*Fomitopsis cajanderi*	None
Gray	*Trametes versicolor*	None
Green	*Omphalotus olivascens*	Iron
Lavender	*Gomphus clavatus*	Iron
Olive	*Boletus mirabilis*	Iron
Orange	*Cortinarius cinnamomeus*	None
Pink	*Chroogomphus vinicolor*	None
Red	*Cortinarius phoeniceus* var. *occidentalis*	Alum
Yellow	*Phaeolus schweinitzii*	None

NOTE: Adapted from Rice and Beebee (2007). Their book contains a more detailed list.

Cortinarius cinnamomeus is one of many species of *Cortinarius* with brightly colored gills used in dyeing fabric.

Fungus Fairs

One of the best ways to learn mushroom names and classification is to visit an exhibit organized by a local mushroom organization. The Resources section (p. 439) describes how to locate the nearest group and find the dates and locations of fungus fairs. A few days before the exhibit, members of the organization range far and wide in search of fungi (to display, not to eat) along with forest detritus, such as pine needles, leaves, and moss, to illustrate different habitats. Mushrooms don't store well, so foraging can't be done too far in advance. Water from spray bottles keeps the collection looking fresh. The night before the exhibit, experienced and knowledgeable folks, often including a professional mycologist from a nearby college, identify what has been collected. All the boletes are placed on one table, the inky caps (mushrooms that self-digest into a black ink) on another table, and so on. Then a finer division is made by genus and species, and the species names are written on 3 × 5 cards and stored with the mushrooms in wax paper bags until the morning of the exhibit. Species are often grouped by habitat to provide a more realistic perspective to the exhibit. Boletes are placed on a bed of pine needles, chanterelles on oak leaves, mushrooms growing on wood will be on or near branches and logs, and so forth. The display should be both realistic and educational.

Mushrooming is a fascinating hobby and an intellectual challenge. It gives you a reason to look forward to rain and makes a walk in the park or woods an exploration. You can photograph, draw, paint, write about, and make dyes from mushrooms. There are mushroom historians and cultural anthropologists, people who document species in a particular location over time, and those whose sole interest is collecting mushrooms for the table. You will meet interesting people and visit out-of-the-way places on forays. Over time you will find your own intriguing aspects of mushrooming.

Agaricus moelleri

Psilocybe cubensis

Clitocybe odora

Leccinum insigne

Tricholoma virgatum

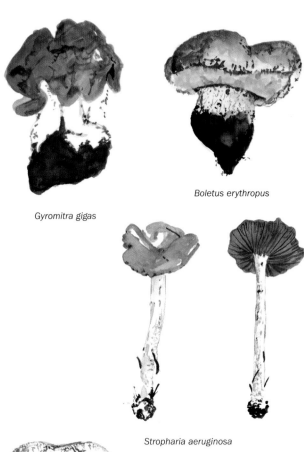

Gyromitra gigas

Boletus erythropus

Stropharia aeruginosa

Scleroderma cepa

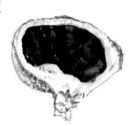

APPENDIX 1

SPORE COLORS OF SOME COMMON GILLED MUSHROOM GENERA

White, Yellow, Green, or Pink Spores

Amanita
Armillaria
Caulorhiza
Chlorophyllum
Clitocybe
Connopus
Cystoderma
Flammulina
Floccularia
Gymnopus
Hygrocybe
Hygrophoropsis

Hygrophorus
Infundibulicybe
Laccaria
Lactarius
Lepiota
Leucoagaricus
Leucocoprinus
Leucopaxillus
Lichenomphalia
Lyophyllum
Marasmiellus
Marasmius

Melanoleuca
Mycena
Neolentinus
Omphalotus
Pleurotus
Russula
Schizophyllum
Strobilurus
Tricholoma
Tricholomopsis
Xeromphalina

Reddish, Salmon, or Pinkish Brown Spores

Entoloma
Phyllotopsis

Pluteus
Volvopluteus

Black Spores

Chroogomphus
Coprinellus
Coprinopsis

Coprinus
Gomphidius
Panaeolus

Parasola
Pseudocoprinus

Brown, Purple-Brown, or Purple-Black Spores

Agaricus
Agrocybe
Crepidotus
Hebeloma
Hypholoma

Inocybe
Leratiomyces
Panaeolus
Paxillus
Pholiota

Psathyrella
Psilocybe
Stropharia

Orange-Brown to Rusty Brown Spores

Bolbitius	*Galerina*	*Tubaria*
Conocybe	*Gymnopilus*	
Cortinarius	*Phaeocollybia*	

APPENDIX 2

SYNONYMS, NAME CHANGES, AND MISAPPLIED NAMES

Other names	Names used in this field guide
Agaricus praeclaresquamosus	*Agaricus moelleri*
Agaricus texensis	*Agaricus deserticola*
Agrocybe semiorbicularis	*Agrocybe pediades*
Albatrellus ellisii	*Scutiger ellisii*
Albatrellus flettii	*Albatrellopsis flettii*
Albatrellus pes-caprae	*Scutiger pes-caprae*
Alboleptonia sericella	*Entoloma sericellum*
Aleuria rhenana	*Sowerbyella rhenana*
Amanita aspera	*Amanita franchetii*
Amanita caesarea	*Amanita jacksonii*
Amanita calyptrata	*Amanita calyptroderma*
Amanita lanei	*Amanita calyptroderma*
Amanita rubescens	*Amanita novinupta*
Armillaria albolanaripes	*Floccularia albolanaripes*
Armillaria caligata	*Tricholoma caligatum*
Armillaria olida	*Tricholoma vernaticum*
Armillaria ostoyae	*Armillaria solidipes*
Armillaria ponderosa	*Tricholoma magnivelare*
Bolbitius vitellinus	*Bolbitius titubans*
Boletus aereus	*Boletus regineus*
Boletus pinophilus	*Boletus rex-veris*
Boletus piperatoides	*Chalciporus piperatoides*
Boletus piperatus	*Chalciporus piperatus*
Boletus satanas	*Boletus eastwoodiae*
Bondarzewia montana	*Bondarzewia mesenterica*
Calvatia utriformis	*Lycoperdon utriforme*
Camarophyllus pratensis	*Hygrocybe pratensis*
Camarophyllus russocoriaceus	*Hygrocybe russocoriacea*
Camarophyllus virgineus	*Hygrocybe virginea*
Cantharellus infundibuliformis	*Craterellus tubaeformis*
Clavaria purpurea	*Alloclavaria purpurea*
Clavaria vermicularis	*Clavaria fragilis*

Other names	Names used in this field guide
Clavariadelphus pistillaris	*Clavariadelphus occidentalis*
Clavicorona pyxidata	*Artomyces pyxidatus*
Clitocybe cyathiformis	*Pseudoclitocybe cyathiformis*
Clitocybe deceptiva	*Clitocybe fragrans*
Clitocybe gibba	*Infundibulicybe gibba*
Clitocybe squamulosa	*Infundibulicybe squamulosa*
Clitocybe subconnexa	*Lepista subconnexa*
Collybia acervata	*Connopus acervatus*
Collybia butyracea	*Rhodocollybia butyracea*
Collybia dryophila	*Gymnopus dryophilus*
Collybia erythropus	*Gymnopus erythropus*
Collybia fuscopurpurea	*Gymnopus villosipes*
Conocybe albipes	*Conocybe apala*
Conocybe lactea	*Conocybe apala*
Coprinus atramentarius	*Coprinopsis atramentaria*
Coprinus auricomus	*Parasola auricoma*
Coprinus disseminatus	*Coprinellus disseminatus*
Coprinus lagopus	*Coprinopsis lagopus*
Coprinus micaceus	*Coprinellus micaceus*
Coprinus plicatilis	*Pseudocoprinus lacteus*
Coprinus stercoreus	*Coprinopsis stercorea*
Cordyceps capitata	*Elaphocordyceps capitata*
Cordyceps ophioglossoides	*Elaphocordyceps ophioglossoides*
Cortinarius alboviolaceus	*Cortinarius griseoviolaceus*
Dentinum repandum	*Hydnum repandum*
Dentinum umbilicatum	*Hydnum umbilicatum*
Dictydium cancellatum	*Cribraria cancellata*
Discina ancilis	*Discina perlata*
Endoptychum depressum	*Agaricus inapertus*
Entoloma madidum	*Entoloma bloxamii*
Floccularia straminea	*Floccularia luteovirens*
Fomitopsis officinalis	*Laricifomes officinalis*
Galerina autumnalis	*Galerina marginata*
Gastropila fragilis	*Calvatia pachyderma*
Gastropila fumosa	*Calvatia fumosa*
Gastropila subcretacea	*Calvatia subcretacea*
Genea intermedia	*Gilkeya compacta*
Gomphus floccosus	*Turbinellus floccosus*
Gomphus kauffmanii	*Turbinellus kauffmanii*
Gymnopilus spectabilis	*Gymnopilus junonius*
Gymnopus androsaceus	*Marasmius androsaceus*
Gyromitra montana	*Gyromitra gigas*
Gyromitra perlata	*Discina perlata*
Handkea lycoperdoides	*Calvatia lycoperdoides*
Handkea subcretacea	*Calvatia subcretacea*
Handkea utriformis	*Lycoperdon utriforme*

Other names	Names used in this field guide
Heterobasidion annosum	*Heterobasidion irregulare*
Heterotextus alpinus	*Guepiniopsis alpina*
Hydnum fuligineoviolaceum	*Sarcodon fuligineoviolaceus*
Hydnum fuscoindicum	*Sarcodon fuscoindicus*
Hydnum imbricatum	*Sarcodon imbricatus*
Hydnum scabrosum	*Sarcodon scabrosus*
Hypholoma aurantiacum	*Leratiomyces ceres*
Hypoxylon thouarsianum	*Annulohypoxylon thouarsianum*
Inocybe jurana	*Inocybe adaequata*
Inocybe pyriodora	*Inocybe fraudans*
Inonotus tomentosus	*Onnia tomentosa*
Lactarius chrysorheus	*Lactarius xanthogalactus*
Lactarius fragilis var. rubidus	*Lactarius rubidus*
Lactarius vinaceorufescens	*Lactarius xanthogalactus*
Laetiporus sulphureus	*Laetiporus conifericola*
Lentinus lepideus	*Neolentinus lepideus*
Lentinus ponderosus	*Neolentinus ponderosus*
Lentinus strigosus	*Panus rudis*
Lepiota cepistipes	*Leucocoprinus cepistipes*
Lepiota clypeolaria	*Lepiota magnispora*
Lepiota cristata	*Lepiota castaneidisca*
Lepiota naucina	*Leucoagaricus leucothites*
Lepiota roseifolia	*Leucoagaricus erythrophaeus*
Lepista brunneocephala	*Clitocybe brunneocephala*
Lepista nuda	*Clitocybe nuda*
Lepista tarda	*Clitocybe tarda*
Leptonia chalybaea	*Entoloma chalybeum*
Leptonia parva	*Entoloma parvum*
Leptonia serrulata	*Entoloma serrulatum*
Leucopaxillus amarus	*Leucopaxillus gentianeus*
Leucopaxillus giganteus	*Clitocybe gigantea*
Longula texensis	*Agaricus deserticola*
Lyophyllum montanum	*Clitocybe glacialis*
Macrolepiota rachodes	*Chlorophyllum rachodes*
Morchella semilibera	*Mitrophora semilibera*
Morganella pyriformis	*Lycoperdon pyriforme*
Mycena alcalina	*Mycena stipata*
Mycena elegantula	*Mycena californiensis*
Mycena griseoviridis	*Mycena nivicola*
Mycena rorida	*Roridomyces roridus*
Mycetinis copelandii	*Marasmius copelandii*
Mycetinis scorodonius	*Marasmius scorodonius*
Naematoloma capnoides	*Hypholoma capnoides*
Naematoloma fasciculare	*Hypholoma fasciculare*
Nolanea hirtipes	*Entoloma hirtipes*
Nolanea holoconiota	*Entoloma holoconiotum*

Other names	Names used in this field guide
Nolanea proxima	Entoloma propinquum
Nolanea sericea	Entoloma sericeum
Nolanea stricta	Entoloma strictius
Omphalina ericetorum	Lichenomphalia umbellifera
Omphalina luteicolor	Chrysomphalina aurantiaca
Panaeolina foenisecii	Panaeolus foenisecii
Panaeolus campanulatus	Panaeolus papilionaceus
Panaeolus retirugis	Panaeolus papilionaceus
Parasola leiocephala	Pseudocoprinus lacteus
Paxillus atrotomentosus	Tapinella atrotomentosa
Paxillus panuoides	Tapinella panuoides
Peziza sylvestris	Peziza arvernensis
Phaeoclavulina abietina	Ramaria abietina
Phaeoclavulina curta	Ramaria myceliosa
Phlogiotis helvelloides	Guepinia helvelloides
Pholiota destruens	Pholiota populnea
Pholiotina filaris	Conocybe filaris
Pisolithus tinctorius	Pisolithus arhizus
Pluteus lutescens	Pluteus romellii
Polyporus badius	Royoporus badius
Polyporus elegans	Polyporus varius
Polyporus hirtus	Jahnoporus hirtus
Porodaedalea pini	Phellinus pini
Prunulus purus	Mycena pura
Psathyrella conopilus	Parasola conopilus
Psathyrella hydrophila	Psathyrella piluliformis
Psilocybe coprophila	Deconica coprophila
Ramaria fennica var. violaceibrunnea	Ramaria violaceibrunnea
Ramariopsis corniculata	Clavulinopsis corniculata
Ramariopsis fusiformis	Clavulinopsis fusiformis
Ramariopsis laeticolor	Clavulinopsis laeticolor
Roridomyces roridus	Mycena rorida
Russula rosacea	Russula sanguinea
Russula silvicola	Russula cremoricolor
Scleroderma geaster	Scleroderma polyrhizum
Scleroderma vulgare	Scleroderma citrinum
Sparassis crispa	Sparassis radicata
Stropharia aurantiaca	Leratiomyces ceres
Stropharia riparia	Leratiomyces percevalii
Tricholoma flavovirens	Tricholoma equestre
Tricholoma pessundatum	Tricholoma muricatum
Tricholoma terreum	Tricholoma myomyces
Tricholoma zelleri	Tricholoma focale
Tricholomopsis fallax	Megacollybia fallax
Tylopilus porphyrosporus	Porphyrellus porphyrosporus
Tylopilus pseudoscaber	Porphyrellus porphyrosporus

Other names	Names used in this field guide
Tyromyces amarus	*Oligoporus amarus*
Tyromyces caesius	*Oligoporus caesius*
Tyromyces leucospongia	*Oligoporus leucospongia*
Vascellum pratense	*Lycoperdon pratense*
Volvariella gloiocephala	*Volvopluteus gloiocephalus*
Volvariella speciosa	*Volvopluteus gloiocephalus*
Xerocomus chrysenteron	*Boletus chrysenteron*
Xerocomus zelleri	*Boletus zelleri*

GLOSSARY

Acrid Sharp or peppery taste.

Adnate gills Gills are broadly attached to the stalk (p. 9).

Adnexed gills Gills that are narrowly attached to the stalk (p. 9).

Amyloid Staining blue, blue-gray, or black in an iodine solution such as Melzer's reagent.

Annulus Ring remaining on the stalk from partial veil tissue.

Apical At the apex.

Appressed Lying flat.

Areolate Divided by cracks and lines.

Ascomycete A common name for a phylum of fungi that produce their spores in a saclike ascus.

Ascus Saclike cell in which spores are produced.

Azonate Without bands of different colors.

Basidiomycete A common name for a phylum of fungi that produce their spores on a basidium.

Basidium Club-shaped microscopic cell on which spores are produced.

Brown rot Wood decay that leaves a brown residue from preferential digestion of cellulose.

Capillitium Sterile threadlike hypha within the spore mass of many gasteroid fungi and slime molds.

Clade A group of related species according to measurable characters, e.g., DNA sequences.

Clavate Shaped like a club.

Cluster To grow closely together.

Conk Firm, woody fruiting body of some polypores growing on trees.

Cuticle Outer surface layer of caps or stalks.

Cystidium Microscopic sterile cell on the cap, gills, or stalk.

Decurrent gills Gills that extend partway down the stalk (p. 9).

Deliquesce To liquefy.

Desiccate To dry.

Dextrinoid Staining reddish brown in an iodine solution such as Melzer's reagent.

Distant Widely spaced.

DNA Molecule consisting of chains of nucleotides and therefore the genetic code; the sequences of nucleotides determine individuality and are used to infer ancestry.

Eccentric Off-center, as in an off-center stalk.

Effused Stretched thin and flat and adhering to the substrate.

Effused-reflexed Partly resupinate and partly shelflike.

Equal In reference to a stalk, one with parallel sides.

Evanescent Fleeting, quickly disappearing.

Farinaceous Mealy, cucumber-like odor.

Fetid Foul smelling.

Fibrillose Composed of small fibers.

Filamentous Threadlike or filament-like.

Floccose Covered with wool-like tufts.

Free gills Gills not at all attached to the stalk (p. 9).

Fusiform Thickest in the middle and tapered at both ends.

Gasteroid fungi Mushrooms such as puffballs that form spores within an internal cavity.

Glabrous Smooth.

Globose Shaped like a globe or sphere.

Glutinous Slimy, especially in wet weather.

Hyaline Transparent, colorless.

Hygrophanous Cap surface changing color, often fading, as it loses moisture.

Hymenium Spore-bearing surface of a mushroom.

Hyphae Microscopic threadlike or tubelike fungal cells.

Hypogeous Underground.

Inamyloid Not changing color in an iodine solution such as Melzer's reagent.

Lacunose Deeply furrowed or pitted; characterized by shallow cavities.

Lamella Gill.

Latex Milky or colorless fluid.

Lignin A complex constituent of wood that binds cellulose fibers.

Lubricious Having a slick or slippery quality, but not slimy.

Marginate gills Gills with edges colored differently from the sides.

Melzer's reagent An iodine solution used by mycologists to characterize spores.

Mycelium A mass of hyphae, the vegetative part of a fungus.

Mycologist Professional who studies fungi.

Mycorrhiza Mutually beneficial symbiosis in which nutrients are exchanged between plant roots and fungal hyphae.

Nucleotide Building block of DNA.

Pallid Lacking color, pale.

Parasitism Deriving nutrition from a host organism.

Partial veil Protective tissue extending from the stalk to the cap margin in many undeveloped mushrooms, sometimes leaving a ring on the stalk or pieces of tissue on the cap margin.

Pendant Drooping, hanging down.

Peridiole Tiny spore-producing chamber.

Peridium Outer layer (rind) of the spore case of puffballs and other gasteroid fungi.

Perithecium Minute flask in which asci are produced.

Phenolic An unpleasant, resinlike odor.

Pileocystidium Cystidium on the cap surface.

Pileus Mushroom cap.

Plane Flat.

Pliant Bending easily without breaking, opposite of rigid.

Pore Tube mouth or opening.

Pruinose Covered with a delicate, flourlike powder.

Pubescent Covered with soft short hairs.

Resupinate Lying flat on the substrate.

Reticulate Resembling a mesh netting.

Rhizomorph Ropelike aggregate of hyphae.

Ring Remains of partial veil tissue on a mushroom stalk, often in the shape of a ring or skirt; annulus.

Saprobe An organism that derives its nourishment from decaying organic matter.

Scaber A tuft of hairs.

Sclerotium A mass of fungal tissue, often underground and tuber-like.

Scrobiculate Marked with pits and spots.

Scurfy Roughened quality, e.g., a scaly deposit on the surface.

Septate Possessing crosswalls.

Serrated Saw-toothed.

Sessile Stalkless.

Slime mold Funguslike, spore-producing organism that forms and moves in an amoeba-like mass.

Spindle-shaped See fusiform.

Sporangium A spore case of a slime mold.

Spores Reproductive unit of a fungus, analogous to the seeds of a plant.

Squamulose Covered with small scales.

Stipe Stalk or stem of a mushroom.

Striate Marked by radiating lines or furrows.

Stroma A hard mass of fungal tissue in or on which fruiting bodies are produced.

Subdistant Fairly widely spaced.

Subglobose Almost spherical.

Substrate The medium, such as wood, leaf litter, or compost, on which an organism grows.

Taxonomy Science of classification.

Toadstool Derogatory term, often applied to a wild mushroom. No technical meaning.

Tomentose Covered with dense wooly hairs.

Truncate Terminating abruptly, as if the end were chopped off.

Umbilicate Having a central depression, as in a human navel.

Umbonate Having an elevated central area or umbo.

Universal veil Membranous tissue covering a young unopened fruiting body, often leaving a volva at the base and warts or patches on the cap.

Vinaceous Colored like burgundy wine.

Viscid Sticky or slimy, at least when moist.

Volva Collarlike or cuplike remains of the universal veil at the base of the stalk.

Wart Piece of the universal veil on the mushroom cap.

White rot Wood decay that leaves a pale residue from the initial digestion of lignin.

Zonate Concentric bands of different colors or gradations of a single color.

RESOURCES

Some National Field Guides

Arora, D. (1986). *Mushrooms Demystified.* Berkeley, CA: Ten Speed Press. (Comprehensive for West Coast mushrooms. Author has a great sense of humor.)

Lincoff, G. H. (1981). *The Audubon Society Field Guide to North American Mushrooms.* New York: Knopf. (Very good key using mushroom shape.)

McKnight, K., and McKnight, V. (1998). *A Field Guide to Mushrooms.* Boston: Houghton Mifflin. (Very nice watercolor paintings.)

Phillips, R. (1991). *Mushrooms of North America.* Boston: Little, Brown. (All color photographs taken in studio.)

Mushroom Cultivation

Stamets, P. (2000). *Growing Gourmet and Medicinal Mushrooms.* 3rd ed. Berkeley, CA: Ten Speed Press.

Stamets, P., and Chilton, J. S. (1984). *The Mushroom Cultivator.* Olympia, WA: Agarikon Press.

General Mushroom Books

Friedman, S. A. (1986). *Celebrating the Wild Mushroom.* New York: Dodd, Mead. (A passionate quest.)

Hall, I. R., Brown, G. T., and Zambonelli, A. (2007). *Taming the Truffle.* Portland, OR: Timber Press. (Everything you want to know about truffles.)

Jay, E., Noble, M., and Hobbs, A. S. (1992). *A Victorian Naturalist.* London: F. Warne and Co. (Beatrix Potter's paintings of fungi.)

Rolfe, R. D., and Rolfe, F. W. (1923/1974). *The Romance of the Fungus World.* New York: Dover Books. (Inexpensive book of mushroom lore.)

Schaechter, E. (1997). *In the Company of Mushrooms.* Cambridge, MA: Harvard University Press. (A biologist's tale.)

Williamson, B. L. (2002). *Reflections on the Fungaloids.* Ottawa: Algrove. (Amazing watercolors by a professional illustrator.)

Societies and Clubs in Western States

Because web addresses and contact people can change, check the website of the North American Mycological Association for the latest information: www.namyco.org/

Periodicals

Fungi, P.O. Box 8, 1925 Highway 175, Richfield, WI 53076-0008. www.fungimag.com/

McIlvainea: Journal of American Amateur Mycology, North American Mycological Association. www.namyco.org/publications/mcil_journal.html

Mushroom, the Journal of Wild Mushrooming. www.mushroomthejournal.com/

Websites

American Association of Poison Control Centers: www.aapcc.org/DNN/ (Poison Help hotline: 1-800-222-1222.)

Fungi Perfecti, LLC: www.fungi.com (Catalog of mushroom products, cultivation equipment, art, and books.)

Index Fungorum: www.indexfungorum.org/Index.htm (A database of current names of fungi and their synonyms.)

Matchmaker: Mushrooms of the Pacific Northwest: www.pfc.cfs.nrcan.gc.ca/biodiversity/matchmaker/index_e.html (Descriptions of more than 2,000 gilled mushrooms of Washington, Oregon, Idaho, and British Columbia. Online entry makes it possible to enter mushroom characteristics and receive a list of possible species that fit the description.)

MushroomExpert: www.mushroomexpert.com/ (Photos, online keys, and descriptions of major groups and species of mushrooms.)

MushroomHobby: California and Beyond: www.mushroomhobby .com/ (Photos of fungi of the western United States and other areas of the world, especially useful for *Cortinarius* species.)

MushroomHunter, Common Names of Wild Mushrooms of North America: www.mushroomhunter.net/common_names.htm

Mushroom Observer: mushroomobserver.org/ (Comparative photo sharing, observations, and open debate on identifications.)

MycoWest: www.mycowest.org/ (Focusing on fungi of New Mexico and the Southwest.)

MykoWeb: http://mykoweb.com/ (Focusing mainly on California fungi, the site contains photographs of more than 800 species plus 200 species descriptions with links to other online descriptions and photos. Contains a glossary, bibliography, recipes, a mushroom blog, and more.)

North American Mycological Association: www.namyco.org/ (The major North American organization of amateur mushrooming, with links to regional, state, and local groups, and the addresses of mycology laboratories and herbaria. Contact information and details about local events, exhibits, resources, and forays are often posted.)

RogersMushrooms: www.rogersmushrooms.com/ (Online photos and descriptions of species in Europe and North America.)

Taylor Lockwood: www.fungiphoto.com (Art, photographs, and videos of mushrooms.)

Western Montana Mycological Association's Mushroom Photo Guide: www.fungaljungal.org/guide/index.htm (Photos, online key, links to mycological articles, detailed accounts of cooking wild mushrooms.)

For users of smartphones and other electronic devices, many applications for mushroom identification are available.

ADDITIONAL CAPTIONS

ACKNOWLEDGMENTS AND ART CREDITS

Many people contributed in large and small ways in the production of this book. We are especially grateful to Karina Perez for providing DNA sequences of unusual specimens, some of which were new reports for the western states. Dimitar Bojantchev was extremely generous with his time, resources, and extensive knowledge of western mushrooms. Dan and Anita Fernandez were gracious hosts on Rocky Mountain forays, as were Walt and Lois Ratcliff on Mendocino visits. Mike Sampson was a steady companion on Sierra forays. We also thank Judy Davis, Barbara Sommer, and Janice Menge for their patience and assistance. Andy Hendrickx and Pam Peterson, Dan Tilley, Dorothy Peters, and Cathy Wolfe alerted us to mushroom sightings. Megan Romberg knitted the cap made with mushroom-dyed wool (p. 421), and Charla and Lani Yakabe drew the picture on the *Ganoderma* conk (p. 349). We thank Elaine Davis for the sketches, and the following contributors of photographs:

DIMITAR BOJANTCHEV *Agaricus arvensis, Amanita pachycolea, Boletus amygdalinus, Boletus eastwoodiae, Clitopilus prunulus, Cortinarius croceus, Cortinarius evernius, Cortinarius laniger, Cortinarius percomis, Cortinarius violaceus, Cortinarius xanthodryophilus, Gymnopilus junonius, Mycena oregonensis, Omphalotus olivascens, Ramaria formosa, Sarcoscypha coccinea, Strobilurus trullisatus, Tricholoma vaccinum.*

TIMOTHY BOOMER (© Timothy Boomer) *Leratiomyces percevalii, Trametes versicolor.*

ALAN ENGLISH *Neolentinus ponderosus.*

NEIL HARDWICK *Bulgaria inquinans, Clitocybe fragrans, Clitocybe nuda, Coprinellus micaceus, Cordyceps militaris, Hygrophoropsis aurantiaca, Hypholoma fasciculare, Lycoperdon pyriforme, Melanoleuca melaleuca, Mycena pura, Phallus impudicus, Psathyrella piluliformis, Trichoglossum hirsutum.*

TAYLOR F. LOCKWOOD (© Taylor F. Lockwood) *Amanita jacksonii, Amanita ocreata, Arcyria denudata, Artomyces pyxidatus, Boletus edulis, Craterellus cornucopioides, Ganoderma lucidum, Gyromitra infula, Hericium erinaceus, Hypomyces lactifluorum, Morchella elata, Russula cremoricolor, Sparassis radicata.*

JAROSLAV MALÝ *Amanita pantherina, Amanita phalloides, Astraeus hygrometricus, Auricularia auricula, Gyromitra esculenta, Pholiota squarrosa, Schizophyllum commune, Tricholomopsis rutilans.*

RAYMOND MCNEIL *Fuligo septica* plasmodium, *Genea arenaria.*

MELISSA H. MORRIS *Amanita novinupta.*

STEPHEN SHARNOFF *Lichenomphalia umbellifera.*

HUGH SMITH (© HUGH SMITH WWW.HUGHSMITH.ORG) *Geastrum fornicatum, Morchella rufobrunnea, Pseudocoprinus lacteus.*

MATTHEW E. SMITH *Hydnangium carneum.*

MATTHEW TRAPPE *Tuber oregonense.*

AMADEJ TRNKOCZY *Ascocoryne sarcoides, Phlebia radiata, Xylaria hypoxylon.*

JULIE WAKELIN *Agaricus augustus, Cystoderma fallax, Laetiporus conifericola.*

RON WOLF *Aleuria aurantia, Amanita vaginata, Amanita velosa, Armillaria mellea, Bolbitius titubans, Boletus zelleri, Calocera cornea, Clavaria fragilis, Clavulinopsis laeticolor, Crepidotus mollis, Discina perlata, Galerina marginata, Geastrum saccatum, Helvella acetabulum, Hygrocybe acutoconica, Hygrocybe psittacina, Hygrophorus eburneus, Lepiota atrodisca, Marasmius plicatulus, Marasmius quercophilus, Mycena adscendens, Peziza repanda, Phaeolus schweinitzii, Polyporus varius, Russula brevipes, Russula cyanoxantha, Xeromphalina campanella* (Cover photograph).

CATHY WOLFE *Agaricus cupreobrunneus, Entoloma sericeum, Lactarius alnicola.*

INDEX

Bold page numbers indicate the main discussion of the species or genus.

ABOUT THE AUTHORS

R. Michael Davis is a professor in the Department of Plant Pathology at the University of California, Davis, where he teaches plant pathology classes as well as courses on mushroom identification, mushroom cultivation, and phylogenetics of fungi. Most of his research is focused on fungal diseases of vegetable and field crops. He has written both popular articles and technical papers on mushrooms, including descriptions of several new species.

Robert Sommer is distinguished professor emeritus at the University of California, Davis, where he chaired departments of Psychology, Environmental Design, Rhetoric and Communication, and Art (not all at the same time). An avid mushroomer and mushroom sketcher, Bob writes the Easy Edibles column for *Mushroom, the Journal of Wild Mushrooming*, and has written and illustrated mushroom articles for *Natural History*, *Horticulture*, *Sierra Heritage*, plus newspapers and mushroom society newsletters.

John Menge is professor emeritus at the University of California, Riverside, in the Department of Plant Pathology, where he taught mycology for 31 years and authored more than 250 publications on fungi. He has been an avid mushroom hunter for over 50 years and has collected and photographed mushrooms and led mushrooming tours all over the world.

Series Design:	Barbara Haines
Design Enhancements:	Beth Hansen
Design Development:	Jane Tenenbaum
Indexer:	Leonard Rosenbaum
Composition:	Bytheway Publishing Services
Text:	9/10.5 Minion
Display:	ITC Franklin Gothic Book and Demi
Prepress:	Embassy Graphics
Printer and Binder:	Imago